This is a welcome addition to t͟ chology. I found it interesting a͟ has a lively readable style to boot.

—Amit Goswami, quantum physicist and author of *Quantum Creativity* and *Quantum Economics*

I was recently working on an article about why Energy Psychology is referred to as such and the integral connection between energy, consciousness, and universal consciousness, when along came Valerie Varan's book *Living in a Quantum Reality*. Synchronicity? Absolutely! This is a remarkable book that offers an in-depth excursion into Valerie's consulting room, and helps us to get in touch with our energy, consciousness, higher self and more. The book explores varieties of our energy from slower to higher vibration, a profoundly important understanding of who we really are. Read this, and read it again and again. Absorb the wisdom and the practical exercises and suggestions for reaching higher levels of vibration and consciousness. This is a book needed for our time.

—Fred P. Gallo, PhD, DCEP, author of *Energy Psychology*, and president of the Association for Comprehensive Energy Psychology

Modern science is beginning to learn that the subtle energy field of the body, also known as the "aura," affects our health and well-being in many ways. It carries the patterns of growth and repair that govern the physical body, and it connects us to the higher spiritual planes involving past lives and the deeper reasons we are here in this lifetime. By working with these deep connections it is possible to achieve healing in ways that are impossible using conventional psychology. Valerie Varan, in her pioneering book *Living in a Quantum Reality*, has taken the important step of showing how these energies can be used in the therapeutic process, and has made a valuable contribution to energy psychology.

—Dr. Claude Swanson, author of *Life Force, the Scientific Basis*

Living in a Quantum Reality brings 21st Century science to the emotional and spiritual challenges of *living* in the 21st Century. Starting with an ambitious, if speculative, synthesis of consciousness studies and quantum physics, the book provides wise guidance for all of us grappling with a world that is in terrifying chaos while teeming with unprecedented opportunities.

—David Feinstein, PhD, author of *The Promise of Energy Psychology* and co-author of *The Energies of Love: Using Energy Medicine to Keep Your Relationship Thriving*

On one level this book offers intriguing information on the intersection between quantum physics and psychology. On another level it challenges therapists and healers to consider the totality, both seen and unseen, of the person in front of them and of the space in between.

—Julie Uhernik RN, LPC, NCC, author of *Using Neuroscience in Trauma Therapy*

As I am reading Valerie's book, I am struck with the exciting possibilities of what it would be like to vibrate with higher energies on a more regular basis. Finally, I can "see" that dissolving my human boundaries doesn't have to feel scary because I might be "losing" myself, but rather that I would be GAINING the Universe and the thrilling possibility of what that might mean. It has helped me realize how my fears are holding me back to "becoming" what I am meant to be. The more I read, the more excited I become and the higher I vibrate!! I can honestly say that for the first time I am understanding what it feels like to "let go" and feel the subtle connectedness to all that is!!

—Judy Ford, LPC, Licensed Professional Counselor

I have, as a clinical social worker doing independent practice twenty-three years, used energy psychology for seventeen years as a mainstay of my work with clients. However, I have always wanted a more in-depth understanding of the way energy actually operates in the body/mind. What Valerie shares about energy makes this possible. I have been astonished at her curiosity into and the scope of her knowledge of energy, and how we can take more advantage of it in the process of growth and healing.

—John Harder, Licensed Clinical Social Worker

LIVING IN A
QUANTUM
REALITY

Using Quantum Physics and Psychology
to Embrace Your Higher Consciousness

VALERIE VARAN

**TURNING
STONE
PRESS**

Cover design by Frame25Productions
Cover art © Neo Edmund and Sander van der Werf / Shutterstock
Interior design by Jane Hagaman

Turning Stone Press
8301 Broadway, Suite 422
San Antonio, TX 78209
www.turningstonepress.com

Library of Congress Control Number available upon request

ISBN 978-1-61852-104-0

10 9 8 7 6 5 4 3 2 1

Printed in the United States of America

With gratitude to all of my teachers, no matter in what plane they reside; to quantum pioneers in every discipline; to my dear clients; and to my husband, family, and friends, who have long supported my soul work. Special thanks to Allison and each of the staff at Turning Stone Press, to Heather Crank for her patience and persistence in illustrating the Consciousness Map, and to Caroline Pincus, my developmental editor, for without her vision, this book would surely have never materialized.

Disclaimer

This book is not intended to dismiss, describe, prescribe, diagnose, cure, or treat any form of mental or physical medical condition or illness. Nor is it intended to serve as psychotherapy, as medical advice, or as a substitute for care under a licensed mental health provider or physician. This book is strictly a guide to the growth of the spiritual soul within each of us. Client case examples are a composite of many clients and are not intended to represent one single client. Any example appearing to do so is with permission or is accidental . . . or, rather, synchronous.

Contents

Introduction

What We Seek from Psychology and Science

On a recent trip to Florida, I felt a pull to sit by a particular young woman, so I did. "You seem bubbly . . . like me!" I smiled, though I don't usually say things like that since on plane rides I tend to sit quietly by the window and read. She responded enthusiastically, and began to tell me about her job in the healthcare field and the masters' program in healthcare administration in which she was enrolled. That sparked my attention, given that I too am in the field. Law of Attraction? Hmmm. As she continued to speak, I would feel inspired to tell her about something of my work, so I would.

Our conversation seemed to naturally flow in the direction of quantum topics, such as energy, coherent fields, the power of love to heal, and so forth. It seemed that any of this information that I would reveal, she would not only gobble up but also instantly generalize to her life, both personal and professional. Then we would talk some more. I would feel intuited to share something else about these topics, she would gobble that up, too, and, again, would completely "get it" and share insights with me that were "downloading" for her.

It was shortly after I had described to her the concept of *archetypal and blueprint fields of energy and consciousness, upon which the material body or form is knitted* that her entire body jerked. She looked stunned, then responded to my inquisitive gaze by saying that she had just felt something like a bolt of lightning come in through the top of her head and run down through her left shoulder.

We laughed as I replied, "Well, I guess you had so many downloads of consciousness as we were talking that your entire body had to be physically adjusted!" It is that instantly that we can take quantum leaps in consciousness.

When we directly experience the mystical and transpersonal, we seem to plunge into the pool of Wholeness that is Source at our core.

Many of us seek scientific information to explain such experiences as spiritual awakenings or metaphysical, near-death, or paranormal phenomena. And we need help in coping with such a multidimensional or "quantum reality" that makes us feel so alien from the majority of those around us.

For us, quantum physics complements neuroscience, psychology, and ancient wisdom teachings to help us understand our atypical experiences. It is no wonder we have felt pulled toward quantum discoveries to help us with the vocabulary we lack to describe the energies we sense and the realms of consciousness that we have traveled.

According to physicist and neuroscience pharmaceutical research chemist William J. Bray (author of *Quantum Physics, Near Death Experiences, Eternal Consciousness, Religion, and the Human Soul*), we have to look to quantum physics to understand our transpersonal experiences in consciousness because it holds the only hard definition for consciousness. And it is quantum physics that studies the nature of nature and reality and has shown the interdependence between consciousness and material reality. Bray himself experienced numerous medically documented "deaths" and then spontaneously came back, reportedly more than thirty different times.

Be that as it may, he reminds us that "there is no solid finalized hypothesis in Quantum Theory that is universally agreed upon." And he emphatically points out that "most of the modern interpretations of Quantum Physics cannot explain" essentially all of the data and outcomes from the most important experiments.

In other words, as Stanislav Grof, MD, agreed in his journal article "Revision and Re-Enchantment of Psychology: Legacy of Half a Century of Consciousness Research," because most of our consciousness data is not explained by our current professional theories, he summarily suggests "a revision of some basic assumptions of modern psychiatry, psychology, and psychotherapy. The proposed changes involve the nature of consciousness and its relationship to matter, dimensions of the human psyche, the roots of emotional and psychosomatic disorders and therapeutic strategy. In the light of the new observations, spirituality appears to be an essential attribute of the human psyche and of existence in general."

Help for the Quantum Self

This book is intended to be a move toward integrating quantum physics and spirituality into the field of psychology to help those who have had such transpersonal experiences and are looking for a way to cope with the larger "unawakened" world.

We could say that this is a quantum psychology self-help book, one that integrates quantum physics into both Western and Eastern psychology and ancient wisdom teachings and one dedicated to helping those who experience a quantum reality. A reality that is not limited to the confines of the material world. A reality composed of multidimensional spheres of consciousness, even ones that send us orbiting beyond this place of spacetime before we anchor back into the dense physical/emotional/mental body that Western psychology knows as the brain/body personality self.

Most of us need practical ways to cope as a conscious soul in this unconscious world, to pick ourselves up when we fall back into ego, to remember our highest purpose, and to "hold steady our light" (an ancient and esoteric way of saying "practice being your soul self" in your daily life) in the midst of all kinds of negative pressure.

Within these pages, I address the most common concerns and struggles of those of us who have had some sort of mystical or spiritual awakening, supernatural experience, or transpersonal/transcendent journey into the "impossible" of multidimensional and higher consciousness.

In the first chapter of the book, using findings from quantum physics, I offer a "CliffsNotes" (brief and more user-friendly version, if you will) explanation of energy and consciousness at large, of the slower or lower frequency vibrations of the ego/personality self, of the higher and faster vibrations of the soul and even higher transpersonal self, and how together they form the *"quantum self."*

Then I lay out a map of the higher and lower consciousness aspects of self, so that we might practically recognize where we are within certain vibrations of consciousness at any given moment. Describing various centers or spheres of consciousness, I point out the varying values, motivations, and aspects of intelligence that shift within us as we switch gears among the multidimensional layers of our overriding consciousness. And it can be very helpful to have the more scientific or technical vocabulary above to express what we are feeling.

Also, as I said, we could use some help with these *shift* struggles, some sharing of experience to navigate this fundamental division we feel inside. So, in subsequent chapters, I write about the multiple "negative side effects" of Wholeness, of having a divine or transpersonal experience, including the ensuing angst, darkness, and despair of coming back to and having to live day by day in this physical reality surrounded by those who don't get us.

Last, in the section on subtle energy imbalances, I address the general problems we face when our "radar" is sensitive enough to pick up subtle energy, including being emotionally overwhelmed and reactive to others' emotional energy, but also the fear we might feel if we find that suddenly we see ghosts, aliens, or other paranormal phenomena. For each of these concerns, I offer some initial actions to take, along with stories of those who have been there.

This book is not intended to dismiss, describe, diagnose, cure, or provide treatment for mental illness or replace any needed professional psychotherapy. I encourage you to seek out such help in person with a licensed professional who is trained in transpersonal psychology and related spiritual issues.

For those of you who are licensed professionals, I hope this book offers you additional information about the transpersonal concerns that are out there and ideas about what kinds of integrative approaches have been helpful for my clients.

In order to protect confidentiality, the client stories within are largely a composite of several people and should not be mistaken for a single client. However, clients have often described their experiences similarly, and that in itself may be helpful information for you. I'll be sharing many examples of how practicing psychotherapy with a worldview that the universe is multidimensional, mysteriously quantum, quite synchronous, and eager to stay in communication with us has helped some of my clients.

For example, one of my clients believed that it was higher consciousness for her to "not take up space" on the planet. Her eco footprint was zilch; she braved the cold without heat and the heat without air conditioning in her tiny rented attic space. She was particularly distressed when she met a man with whom she enjoyed a great chemistry but who made significant money and drove an expensive car. She was actually embarrassed for her friends or coworkers to see that car. But using the Consciousness Map, she was able to see how she was limiting herself so much in her life that she couldn't achieve what her soul was aching to create on this plane through her ego, mind, and body.

Through our work together—which helped her focus on balance by shoring up the personality self so that the higher self would have a strong vehicle of expression—she decided to allow herself to try the new relationship, to buy a small home for herself where she planted an organic garden, and ultimately to go on to do international consulting on organic farming.

When we're having these kinds of experiences, we are relieved to read the research that backs up our sense that we are not crazy. It helps to know that physicists and other professionals are studying and documenting their own similar intuitive and psychic abilities, as well as transpersonal and even paranormal experiences that give rise to unorthodox belief systems and the more quantum-edge worldview. The evolving field of quantum physics, though hotly debated even from within the field, is leading us to a more thorough and technical understanding of the self and of all reality as multidimensional layers of energy and consciousness than is the field of psychology.

My Story

I didn't plan on being a therapist, but as many of you likely already know, the Universe does have its way of getting our attention, if we only listen. When we do, the synchronicities become ridiculously evident in our life. When we don't, it seems as if life is nothing but a constant battle.

For me, the Universe started turning up the heat on me to change careers in the mid-1990s when I was still a business development manager in the environmental engineering field.

One day, I had headed off to my usual unhealthy fast food lunch and had just returned to the fourth story building overlooking the Atlantic Ocean where I had my office. I was in the elevator when I noticed a woman staring at me. The kind of stare that burns holes in the side of your head. When it didn't subside, I turned to her, "Excuse me, can I help you?" And the conversation went something like this:

"Are you Dr. Smith?" she inquired.

"No . . . I'm sorry," I answered, thinking that would be that.

"You're not Dr. Smith?"

"No." I smiled.

She paused, continued to stare, then asked again, "Are you sure you aren't Dr. Smith, the psychologist?"

"Pretty sure." I smiled again, laughing to myself and thinking how slow the elevator seemed this day.

"Are you some kind of psychologist, psychiatrist, counselor, or something?" She sounded incredulous and as if she were really trying hard to figure out some incongruence she perceived.

"No . . ." I smiled yet again, then added "I've been thinking about becoming a psychologist, but I'm not one yet . . ." my voice tapered off, a bit wistfully perhaps.

This actually, believe it or not, went on for a little while longer, until she finally exclaimed as she got off on the third floor, "Well! I can't believe you're not a psychologist or something!! You just have that aura!"

What were the chances of that? And who in south Florida in those days used the word "aura"?! Not anybody I knew. I got off on the top floor and hurried to tell my colleague of the strange and unlikely exchange that had just taken place.

This was one of the oddest, but not the last, of the innumerable synchronicities that were about to unfold as the Universe prodded me in this direction.

Despite such synchronicities (e.g., hearing repeated messages, finding improbable transitional career work, the phone ringing with an offer of financial support from quite unexpected sources at the exact moment I finished yelling aloud at God for it in a moment of panic) that arose as I was moving into the field of psychology, there did come a time during graduate school when it seemed as if all help from the Universe had frozen. I plummeted into a Dark Night of the Soul, though at the time I had no name for the unfamiliar feeling of despair that arose from the sense of being disconnected from, even deserted by, God.

Upon listening to me describe what I was feeling, a Native American professor (yep, meeting her seemed guided, too) handed me some sixteenth-century writings by St. John of the Cross. I came to learn from St. John that the Dark Night was considered to be a milestone in each disciple's life, a time when the tangible sense of Divine connection seems severed and results inevitably in a spiritual or transpersonal depression.

To clients going through it, I would later explain that it is as if Father has been holding us up on our bike while we are learning to ride, then suddenly surprises us by letting go. The training wheels come off, so to speak. The lesson may be for us to learn how we will handle things spiritually on our own; it is indeed a gift for us to learn the reach of our own creative power, rather than overly depend on Spirit to rescue us all of the time. But it seems to come on when our ego is determined to be in charge of our spiritual path or intervenes while we think we are following spiritual purpose. It is as if Father says, "Have fun with that! See what you can do if you will." A lesson I would have to learn more than once.

As the lesson also relates to the Law of Attraction, I learned that when we are aligned with God, Universe, Higher Self, Soul, whatever name we give that higher frequency of nonlocal Source or Love energy that unifies us all as one, it is then that we participate in the Law of Attraction. When we are in ego, we are left to work really hard, because we are plunged into the slower frequencies of electrical energy.

Ironically, perhaps, and certainly cliché, it is when I surrendered my ego's will for God's will that the connection I had with Spirit returned as quickly as it had disappeared. But of course that kind of surrender is always easier said than done.

In private practice for more than nine years now, I have continued to broaden my understanding of all of those experiences that Western psychology says are impossible, but that quantum physics, transpersonal psychology, and consciousness researchers document repeatedly. I have read as much quantum physics as I can, studied qigong with various teachers and a grandmaster, learned multiple forms of energy and distant healing, and continued to read other professionals' findings about the multidimensional states of being that we seem to access by way of all kinds of unique paths.

The appreciation that I hold for each of our quantum pioneers, from Einstein, Bohm, Bohr, and Penfield to Swanson, Goswami, Oschman, Pribram, Hunt, Pert, Becker, Tart, Grof, Tiller, Bray, and others too many to name, is beyond my ability to put into words. I remember the first time I came face to face with some of these physicists at the International Society for the Study of Subtle Energies and Energy Medicine (ISSSEEM) annual conference. Tears rolled down my cheeks as I listened to these mystically inclined scientists; I felt as if I were finally home. Their words melted into my being as an affirmation to all I knew as a soul. Maybe it was because I really was such a scientist during the time of Atlantis, as a psychic woman revealed to me one day as she was leaving the conference. What I can say is that each of their presentations resonated with me at the deepest of levels.

While in my twenties I couldn't understand my insatiable impulse to read all I could about three seemingly unrelated fields. Thanks to their groundbreaking work, I now know why I was called to integrate quantum physics with psychology and ancient wisdom. It was to help those who have experienced transpersonal events develop the skills they need to live from soul consciousness within a world that has not yet awakened to the ways of higher love.

Chapter One

Introduction to Energy, Consciousness, and Your Quantum Self

Envisioning Yourself as Energy and Consciousness

You know you are energy. You feel it. In yourself. In the people and the things around you. If you didn't, you probably wouldn't be reading this book.

You know you are made of consciousness. You feel how this energy is brimming with intelligent awareness, with an aliveness that is more than what you've been taught in school.

For those of you who would really like to understand the "CliffsNotes" version of what quantum physicists are telling us about this energy, this consciousness out of which we are made, read on! Here is where we will simply define both energy and consciousness and differentiate between the various electrical, magnetic, and even torsion fields or layers of our being.

The lower and the higher self talked about with some scientific vocabulary at last!

Energy as Power and Frequency of Vibration

We are made of **energy,** the vivifying **power** at our core that enlivens every atom, cell, and material body.

This energy or power **pulses** and vibrates as it moves in and out and throughout the body. We feel it in our heartbeat. We see it in our brain wave pattern. We know it by the sense of life and conscious awareness that it brings.

As this power pulses, it sends out energy **waves** of its **vibration.** Picture squiggly lines that appear to snake upward and dip downward repeatedly, in a repetitive pattern of cyclic peaks and valleys.

As these vibrations travel *outwardly in all directions from their central pulse,* they form *patterns, like concentric rings or circles.* We see these patterns really well when drops of a gentle rain fall upon a quiet lake or when a boat speeds by and sends out a wake behind it.

Each of these energy vibrations travels in its own unique style and pace (i.e., has its own energetic blueprint, signature or fingerprint), kind of like how some people prefer rap and others resonate to opera.

When the peaks and valleys match up, we say the energies or waves are in sync; that is, they couple and **resonate** with one another. They can then become a **coherent** set of energy, a field that is more organized as a harmonious pattern, and therefore more powerful. Once waves resonate with one another, information is shared through a process called **entrainment,** where there is a mutual vibratory effect, and they become locked in phase with one another. This is why we can feel like another person is "on our same wavelength" and why people in a close relationship often feel telepathic with each other. This is also a basis of energy healing and the power of intention.

The rate at which these waves of energy move is called their **frequency.** The slower/lower their frequency or speed, the more material and solid the pattern these vibrations form for our eyes. The faster/higher their frequency, the more subtle or invisible is the pattern that we perceive, as with air or ether.

As you'll soon see, the many and diverse waves of vibration that comprise the self are constantly traveling, spiraling in all directions, and crisscrossing one another in all facets of the body inside and out. Picture children running and spinning all over the playground . . . after having eaten all kinds of sugar!

For now, let's divide these energies into two categories, those vibrations that are slower than or at the speed of light and those that are faster.

The Slower Energies of Electromagnetism: The Lower Self within Time and Space

Electromagnetic energies are those that travel at or slower than the speed of light. In fact, they are an array of light, as we know light, ranging in color depending on their bandwidth of speed or frequency.

We see within a rainbow only the spectrum of light that is visible to human eyes. That is, red, orange, yellow, green, blue, indigo, and violet. The red side of the rainbow is the slower moving light, the lower frequencies. The violet side of the rainbow is the faster or higher frequency light.

Each shade of color is also associated with a sound or musical note that is related to its frequency. For example, according to famed musician and sound healer Jonathan Goldman, very dark red corresponds to the note of G, darkish red to G#, reddish orange to A, and light or yellowish green to B. Bluish green is associated with the note of C, indigo to D, dark violet to E, and very dark violet to F.

This bandwidth of light or electromagnetic energy that is visible to our human eyes and audible to our ears is tiny, like the eye of a needle. In our current understanding, at least 96 percent of the larger reality is unseen by us and is still awaiting our detection and definition.

Just as with color, most of the sounds played in the universe are far beyond the range that our human ears can pick up and detect. P. M. H. Atwater tells us, "Shaped like a torus donut, the heart field busily converts one form of energy to another as it generates an infinite number of harmonic waves. These harmonics run throughout all bodily systems and are so sensitive that they react to conditions four to five minutes before actual occurrence. This futuristic awareness tells the heart if what's coming is positive or negative so it can prepare. First the heart feels the coming event, then the brain is aware of it, then the eye sees it."

Within us, our electromagnetic matrix of light energy seems to be made up of highways of electrical fields intersecting with highways of magnetic fields, *vibrating perpendicularly to each other and radiating outward as orbs or spheres of light.*

Together, and blended with even higher energies, they form the field of energy that is our colorful **aura,** the one that radiates from us and that can be seen by those with extrasensory clairvoyant sight. The aura is *the aurora borealis of our own earthly body, and it changes with every thought we have and with every emotion we send out.*

Electrical versus Magnetic

Electrical energies are the slowest, lowest frequency energies used to communicate information at the most dense physical layer of our body. In other words, electrical energy is *most associated with the particles or things* out of which we are made, like particles of electrons carrying a negative charge that gravitate toward protons carrying a positive charge within the nucleus of an atom.

When electrons travel along highways in our tissues, they are using electrical energy; they are actually generating electricity in the form of electrical currents. These electrical currents serve as communication signals to the rest of the body so that each part knows what to do.

According to research by the Institute of HeartMath, the electrical signals used by our heart are about sixty times the amplitude of the brain's. No wonder we're pulled to follow our heart!

On the other hand, ***magnetic*** *energies seem to be mostly associated with photons, the smallest packets of light*, that appear to *blink on and off* within our body. Yet they seem to naturally arise and accompany the electrical energy of the body's material particles, unless forced apart.

To better see the magnetic field in your mind's eye, think of the Earth's magnetic field, which extends from its core and connects with the incoming energies of the sun. Pictures of it show the field coming into the north pole as we know it, coming out of the south pole, and circling back up to the north on all sides.

In pictures of the biomagnetic field of the human body, we see a similar **toroidal** shape, with energies streaming through the crown of the head and the feet area and circling along the sides, all around us like a deep donut. When researchers are studying energy healers, they tend to measure how strongly the biomagnetic field emanates from the hands, heart, and around the body (up to fifteen feet!) of the healer.

Significantly, according to researchers, our biomagnetic field tells us more about what is going on inside our body than any electrical measurement. When we get an MRI scan, it uses magnetic fields and magnetic resonance (i.e., peaks and valleys line up) to align the protons of our hydrogen atoms in order to detect minute differences and changes within the structures and tissues of the body.

Within the human body, our heart seems to be the most significant transmitter/receiver of core magnetic energy. It has been shown to generate magnetic fields of at least five thousand times that of the brain (within which the pineal gland is said to be the strongest). Some estimate it to be up to a million times the magnetic field of the brain, especially during states of advanced energy healing.

We do know that the overall shape of our body's *biomagnetic field depends mostly on the energetic current emanating from our heart and how coherent or organized* it is. Any frustrations or negative emotions we experience reduce our heart coherence, while positive and loving states increase it. In a state of optimal coherence,

According to the Institute of HeartMath, when we intentionally send out positive energy, it changes the information in the electromagnetic field, to which the entire body adapts. When others are feeling our emotional energy, they are feeling our auric or electromagnetic field, which can be measured at several feet away from our body. When we touch another or are in close range to one another, there is an exchange of electromagnetic energy led by the heart.

we are experiencing unconditional love and gratitude. This is the state we seek when we are learning to meditate and to generate healing energy.

Yet *every movement* that takes place within the body sends out *its own biomagnetic signature or fingerprint,* which is picked up and responded to by all other parts of the body, as one **tensegrity** system (a term borrowed from architecture that refers to how a movement in one area of a structure is automatically compensated for by the rest of the structure).

Combined as one, our electromagnetic field is naturally a coherent set of light energy. Our heart's electromagnetic signals can affect or entrain the brain waves of another person, and we can thus become synchronized to one another. Furthermore, as we increase psychophysiological coherence with one another, we become even more sensitive to the electromagnetic signals of those around us.

The higher our heart coherence, the more likely our brain will entrain to our partner's heart signal, even when we are not touching one another. We will feel more intuitive when we are listening to them.

Through sustained sincere heart states of love and appreciation, the heart in a state of coherence entrains the electrical activity of our own brain, notably the alpha and lower frequency brain waves. It is in this way that when we practice states of positive emotion, especially the emotions of love and gratitude, we can cohere our heart's field, alter our brain activity, improve cognitive performance and decision making, change perceptual awareness, and enhance our immune system.

These electromagnetic light signals are used by our heart, the brain, and our entire nervous system to catalyze and orchestrate all of the biochemical and physiological activity in the body, even at a distance. These signals, exchanged by every cell, are the impetus for the more physical basis of the mind-body connection; they provide the pathway we can use to modify our neurotransmitters, change the wave signals in our brain, and transform our body and its behaviors.

So it appears that the heart motions or entrains the brain, which signals the rest of the body to play along with it, as three harmonious notes in a single chord simultaneously. And a change in one note creates a corresponding change in the other two.

This **electromagnetic self** we will call the **lower self,** because from it is comprised our three lowest, slowest vibrational layers. We can consider it to be the self where resides our physical body, our thoughts, and our emotions. Most of us call this aspect of ourselves our **personality** or **personal self.** Some refer to this self as the **ego.** We will use these terms interchangeably.

This lower self seems to contain *the least amount of our individual soul energy.* That is, at the building-block level of atoms and molecules, it may be that our *magnetic energy (through our biophotons) is the faintest hint of our individual soul essence.* And electrical energy comprises the most physical particle or body nature of us, through which our soul energy moves and expresses itself in a limited or containerized form.

But as we follow the layers of our being from body toward soul, energy seems to transition in form from electrical energy to magnetic, electromagnetic, photonic, and eventually into torsion fields of light.

The Superluminal Energies of Torsion: The Higher Self beyond Time and Space

Toward the higher vibrational pole of our even broader energy spectrum are **torsion fields.** Torsional energies are those that seem to be associated with the left-handed and right-handed spin motion of waves and "generate a twisting of space," according to MIT physicist Claude Swanson. Their signals appear to travel faster than the speed of light, so fast that they take no time at all!

These torsion fields have been documented to alter mass or matter, even across great distances (that is, **nonlocally**), and work forward as well as backward in time. As such, they defy both space and time.

At this level of energy, it is *as if all is one present moment.* It is like there is some unseen conductor who is orchestrating a beautiful symphony, all at once and across great distances. And though the trombonists may not see the drummers, all musicians are playing in harmonic synchrony, in tune with one another across the echoes of space and time. Mindboggling for sure!

Yet these torsion fields are currently being studied by scientists to learn how they might be responsible for subtle energy and phenomena such as telepathy or group mind, intuition, the power of intention, remote viewing, energy healing at a distance, and other "magical" types of psychic abilities.

Clearly, at these levels of energy resides way *more of our soul energy,* our more subtle energy light body radiating waves of power outwardly in all directions, much like the physical sun. As our own central sun, the soul is the strongly coherent core power of energy within us that shapes the electrons and molecules into the pattern for our outer physical body. It is this torsion or subtle energy aspect of ourselves that *exists first nonlocally beyond time and space* and is the part of us that *continues after death of the physical body.* For as we know, energy cannot be created or destroyed, only transformed from one form to another.

> Our soul is a unique "solar power," a unique energetic fingerprint or blueprint pattern. A one-of-a-kind part within the bigger whole. In fact, scientists tell us that no matter how far we go down the quantum rabbit hole, each part is still a unique blend of energy. For example, look out your window and find a tree. That tree is a unique energetic fingerprint. But each leaf on that tree is also a unique energetic fingerprint. On one leaf, each cell is a unique energetic fingerprint. Within one cell, each atom, each proton, and so forth.

Many of us have the sensitivities to feel the presence of this **higher self.** In psychology, we call it the multidimensional **transpersonal self.** In physics and energy medicine, it is called the **subtle energy self.** In religious circles, it is commonly referred to as the **soul self, spiritual self,** or simply **Self.** This is our faster layer of energy that seems to be comprised of torsional energy vibration and beyond.

In Our Wholeness, We Are a Multidimensional and Holographic Matrix of Energy

There are apparently *infinite layers to the matrix of quantum energy* that comprise our being, both visible and invisible.

In our wholeness, we are nonlocal and **multidimensional** in our nature—electrical, magnetic, photonic, torsional, and an even greater number of unique dimensional layers of intelligent energy than we can ever measure—all interwoven and moving together beyond spacetime as one whole or integrated field, just like within white light, when the red, orange, yellow, green, blue, indigo, and violet layers or frequency bands coexist simultaneously as one integrated and overarching field of energy.

It does seem that some great initial Source of power radiates pure energy, which is instantaneously transformed and transduced *from higher and more formless states* of energy into *slower and lower (increasingly physical) forms and structures* of energy as it makes its way to us eventually as torsion, as coherent laser-like light, as a blueprint field of biophotons, as magnetism, and then electricity. In other words, the originating layer or Source projects another layer into existence or reality, which creates another layer, and so forth.

This is how we come to have this more dense electromagnetic light body, the personal self we sense as "I." As it turns out, we literally are "light from light." (Remember $E=MC^2$? Where energy is matter or matter is energy depending on the speed or frequency of its light?!)

It is this aspect of projection and reflection that we can call the **holographic self.** For it is in this way that we are created in the multidimensional light image of Source. A hologram is a three-dimensional light image (think of the projected Princess Leia in *Star Wars*), a coherent energy blueprint that is formed from unique patterns of energy waves as they merge together. The key for us to truly realize, as a multidimensional or quantum self, is that *in a hologram, the information for the whole is contained within each part.*

Stanford neurosurgeon Karl Pribram indeed found that the brain works as a quantum holographic processor of all of these amazing energies.

From his and others' work, we have discovered that mind and memories are evidently not produced by or localized in the brain. Instead, the *information or memory bank for the whole self* is orchestrated nonlocally and is holographically embedded—and therefore *known—within each part and cell.* This is apparently a technical basis for why buried memories can surface during a massage and why body-based and energy therapies, such as tapping and imagery, work to release traumas and pain from our tissues, regardless of whether the trauma is of this present life or some other dimension past or future (yet consider that all lives are really simultaneous lives at the level of the soul and are interconnected through the spacetime point of our present moment awareness).

Photons, the little packets or quanta of coherently organized light usually associated with magnetic energy, actually *exist at multiple frequency levels* and, as such, project quite an unimaginable array of spatial information. Photonic fields truly serve as a significant part of our holographic template, upon which all of our cells depend for their growth. Our biophotons, the photons within the sphere of our electromagnetic body, couple or are "in phase" with our DNA, allowing for the transfer of energy and information (i.e., entrainment) from all of these realms and into the physical body and, as such, are great mediators for our multidimensional consciousness (as we will discuss in the next section).

To summarize, *all aspects of the quantum self, being multidimensional and holographic, exist simultaneously yet at their own particular bandwidth of vibration, just as all colors of a rainbow exist simultaneously within the wholeness we see as white light.*

The personality self seems to be a dim reflection of the soul self, which in turn is a "light" image projected by the even higher self, and so forth. All is power or energy in its various forms and structures.

All layers of this infinite self are there awaiting our perception. And each forms a note, in the song of our soul, in the symphony of creation. Since scientists are now beginning to conclude that *all matter* has a holographic nature, a complex pattern of merging energy frequencies coherently communicating and exchanging light radiation in the past, present, and future simultaneously, *all matter is alive, singing, and participating in this symphony.* We truly are interconnected with All That Is.

Consciousness and Its Qualities, the Information that It Carries

Now, let's get back to simplifying.

Energy waves carry information.

We know this is true because we are using the Internet and computer hardware and software to capture this information and to transform it into things we can see and hear. Every time we turn on the television, listen to the radio, or use any kind of technological device, we are using information carried by energy.

As we have already mentioned, *the information that energy carries is associated with a signature shape or pattern, a blueprint* that serves to guide physical construction, transmission, or expression.

The information within these energy waves communicates a message. It is as if this *information is the code for life itself.*

It is brimming with intelligence, awareness, and, perhaps most fortunately for us, adaptability, the capacity to evolve or reshape itself.

Since energy inherently carries such intelligent patterns of information, we are going to call it consciousness. In other words, let's think of **consciousness** as a field, matrix, or sphere of energy carrying some kind of patterned and holographically coded information. ***Photons*** *are the **primary intermediary,** carrying information from consciousness outside time and space to consciousness within time and space,* translating the code of life into the more magnetic and, in turn, electric frequency bands of life energy that comprise our physical realm.

In this way, everything literally "contains" consciousness (these multidimensional energy fields infinitely carrying patterns of information), and, as such, everything possesses some level of active intelligence. This is what makes each "thing"— whether it is an electron, a tissue cell, a tissue box, a plant, an animal, a human, an angel, or any other being—adaptable to and in communication with its environment.

Technically speaking, e*very "thing" is made of consciousness. As such, energy carrying patterns of information as consciousness is primary and fundamental.* It exists first. It then takes shape as some *thing,* according to the blueprint at its core essence, a coherent field of in-formed energy.

If we consider Swanson's description for consciousness, we come to realize that the more information present, the more self-generating and sustaining the energy field or pattern in matter, then the more "conscious" is the field or sphere of energy. He defines *consciousness as the ability to "steer the phase" of impinging background energy,* which enables a conscious entity to create and maintain its own energy pattern (i.e., as a coherent field of photons and torsion), and to impose this energy pattern on the outside environment and even on distant matter (as an expression of free will). He says that the *more inanimate or less conscious an object, the more its particles are determined by the environment.* Notice that some crystals (even those in a technological device) can have more of an effect on a human being than a human can have on a crystal, so it can be quite debatable—which is more conscious, according to Swanson's definition!

Nonetheless, all that we see as matter is really energy that on some level is conscious and capable of interacting with and responding to its environment.

The information, brought forth by any particular energy, is perceived by us as some type of **quality.**

We might perceive an energy or sphere of consciousness as hot or cold, damp or dry, moving or still. It could feel homogeneous in its sensation, pretty similar throughout. Or it could feel as if it has rough patches or holes. It could feel so stuck or thick that it has a distinct shape with edges. Or it might feel fluid or thin, light, and airy. It generally has a color to it. At times, we sense its sound. There may be a taste or fragrance we detect.

For instance, within our human body, we can perceive our heart space as tense, contracted, a muddy red or dull gray, thick and heavy like a wall, cold and still, or with holes of hurts from events past. Or we could feel our heart as light, warm, moving, expansive, open, and radiating green light to all who pass. When we are trying to describe physical symptoms to our doctors, it is these types of qualities and sensations of energy that we are attempting to put into words.

It seems there are countless qualities of information carried by energy, as consciousness, limited only by our ability to perceive.

We could say that *consciousness is a multidimensional energy field of qualities,* each *quality representing information from a unique frequency domain that we perceive as or through some sense.* Together, these qualities *give us a sense of presence or experience* in the form of our five physical senses, as well as our emotions, thoughts, and creative intuitions.

> There is an ancient teaching that because there is sound, we have an ear. Because there is mind, we have a brain. And because there is love, we have a heart.

So it is no wonder that the most awe-inspiring experiences are beyond words, because they are comprised of all of these qualities coming in from all of these dimensions, both within time and space, and beyond. Yet we are limited to the physical senses to detect and describe them.

How We "Download" Consciousness

Some type of energy receiver and transmitter—some type of antenna—is used to pick up, detect, translate, and communicate the information within energy, within various layers of our consciousness.

When it comes to us humans, scientists have shown that every hair, every cell, every atom in the body serves as an antenna for energy reception and transmission. Imagine that!

Each of us is a channel for energy reception and transmission. Each of us serves as an antenna for higher energies to be converted or transduced into slower, more material, electro-magnetic forces.

As a channel, we ground the higher energies to the lower. We are a rainbow bridge of many colors. And each of us has the opportunity to bridge, to transmit, to actually bring heaven, the higher, down into the lower, this material realm in which we currently live.

I have come to see the *heart as the primary antenna for the life energy* coming in from the universe in its various forms. I believe the heart is intended *to entrain us to the higher* realms and *to bridge us* from the nonphysical center of our being into the physical realms. It is the heart that brings to us our sense of direction and higher purpose. It *delegates to the brain* the job of relaying the information it receives from the universe to the rest of the body by way of the slower, more linear language of the physical, which is electrical. In other words, *the energy is eventually converted from those superluminal energies for which we have no name but God into torsional, photonic, magnetic, and electrical energy.*

I see the *physical brain (especially the pineal gland) as an antenna secondary to the heart.* Yet the brain is the primary antenna, receiver/transmitter, for the physical body and its slower electrical signaling processes. I also see the brain functioning as on old-fashioned call switchboard, which lights up to indicate where calls are taking place in the body. But the brain should not be confused as the originating source for all energy signals carrying information as consciousness, as a switchboard would not be confused with a power plant.

Our experiences in consciousness, though they reside or originate beyond and are not produced by the physical brain (mathematically shown by physicist/pharmaceutical chemist W. J. Bray and experimentally documented by neurophysiologist Wilder Penfield), are certainly registered, monitored, channeled, and filtered by the brain in whatever capacity it exists at the time (e.g., awake, brain dead, on anesthesia, in a coma, in a fog, on drugs, in the dream state, in deep states of meditation). The limits of our brain do hinder and shape our ability to consciously perceive and recall our experiences of consciousness, especially if we are not ready to handle them.

It seems as though the information and the energies of the higher self come in relatively whole, thanks to the right brain antennae; then they are transformed into increasingly useable forms as linear thoughts, thanks to the left brain antennae. Though in reality, all parts of the holographic brain are involved, and its neurons are like musicians on stage waiting to play their particular instruments in time. And it seems as though if something happens to any of the musicians (like when they get sick or die), the other musicians work to pick up any slack in tempo or chord.

However, it is significant for us to grasp the modern-day scientific assertion that *consciousness is a field of intelligent energy that exists first,* before any of us are born. And, as we are about ready to grasp, the quantum field of consciousness as a whole is the background energy or **quantum vacuum** out of which all creation appears to emerge (including the brain!) as it steps down into its lower and slower vibrational forms.

The Creative Power of the Organized, Coherent Acts of Consciousness that We Call Intention

To this point, we have come to see that we are made of multiple levels or layers of energy carrying information, which we simply call consciousness.

This *matrix of consciousness, in its wholeness, is multidimensional and holographic, existing all at once as limited within time and space and unlimited beyond time and space.* For us, it exists as a *series of spheres of consciousness within wider and broader spheres of consciousness.*

This matrix of consciousness is the background and foreground through which we operate. In the *background of space,* this consciousness is unlimited in its potential and power for creativity. In the *foreground lies creation,* condensed into time and space, appearing as matter and brought forth by coherent acts of consciousness.

Let me explain how it is that we are able to join in acts of creation by pointing out the difference in power between energy that is scattered or random and energy that is tightly organized and in a pattern, like a laser or a spiral. Physicists call such organized energy "coherent fields of energy or consciousness." It is these that they have studied in their intention and energy-healing experiments—and these that we will explain further below.

Energy can be more or less organized in its pattern or shape. **Scattered** energy waves we could depict as a scatter plot diagram, a bunch of random marks inked without thought all over a page. These are energies that are not in phase with one another; they are not technically resonating together.

Whereas *organized energy waves tend to vibrate in step with one another, to be in synch as shape and form; scientists call it **coherent** energy. When it is very dense, it appears to us to be material,* to be of substance. But even energy that is not material to our eyes can be quite organized. Take a laser, for instance. It is structured in a tightly spiraled fashion that is coherent energy. Notice the shape and immense power of hurricanes and tornados when energy coheres into those kinds of strong weather patterns.

Coherent fields are thought to be responsible for producing the holograms described earlier and are *what we create when we effectively meditate, use imagery, act in faith, and set intention.* With *sustained focus,* we sweep into power (like a hurricane) a great many of the energy fields to which we have access and *organize these fields of consciousness in the direction of our intention.* Lynne McTaggart has led a movement of intention experiments that anyone can join. I recommend that you check out her website and books on the subject of intention at www.lynnemctaggart.com.

Some of my favorite experiments by physicists on intention involve *meditators who have been shown to influence even random event generator (REG) devices in the direction of their intent.* The significance of this is that REGs, by definition, are machines programmed to move or put out data randomly. In one experiment, however, a roving REG was put into a small rectangular room that contained an empty cage at one end. As predicted, the REG moved around randomly. Until the next part of the experiment.

Chicks that had just hatched were led to believe that the REG was their mother. They imprinted on the REG. When they were then put into the cage within the room where the REG was roaming, the REG began to hover at the side of the room where the baby chicks were in the cage! We can theorize that the baby chicks thought that the REG was their mom and sent some kind of intention or coherent field of "chick love" toward the REG. *(Now if baby chicks can trigger the Law of Attraction with the power of their love, surely we can!)*

Scientists have *correlated coherent energy fields in the body with health and well-being*. Such scientists say that effective energy healers work from coherent fields of higher magnetic and torsion energies; their blueprint energy can affect the ones being healed, who are open to their influence, because the healers' faster and broader vibrational consciousness can entrain and cohere the scattered and slower more disorganized energies of their clients.

Many energy healers say that what they do is tap into the *great power of unconditional love, higher love* that pours down from the field of consciousness as a whole. And they notice that the more they dissolve their personal sense of consciousness into the ocean of consciousness that is cosmic love energy, the more effective the healing that takes place in the body of the ones being healed. It seems that *unconditional cosmic love may be the most coherent energy of them all, perhaps the field of quantum consciousness itself,* cohering us all together, in spiral cyclic fashion, as one.

On the other hand, *scattered energy in the body has been correlated with disease, physical and psychological.* Most of us are quite random in our thinking; therefore our activity, and thus our emotional states, tends to be reactive and fearful instead of intentional and loving. This has a definite impact upon the body. Chronic stress is believed to scatter the fields of our physical body, causing it to become inflamed and to actually fall apart. When we have not yet learned to meditate, to discipline the mind, to focus our thoughts in a constructive, loving, and coherent pattern, we often feel "scattered," because evidently it is quite literally so.

It appears that *people who see the Law of Attraction work*

*synchronously in their lives are those who have learned to create
or align with coherent fields of energy and higher consciousness,
those nonlocal fields that have the
actual power to organize even distant matter.* Those who are usually scattered in their energy fields
are not generating coherent fields
with the power they seek to influence matter. They lack the power
they need (at the level of quantum
physics) to consciously create.

I have come to see that at the physical level of electricity, negatively charged particles seek out positively charged particles. Yet at the levels of our soul, we are attracted to those who are similar in vibration (universal Law of Attraction) and who resonate with our intended purpose. It is these fields that group us together as organs in a higher body.

So perhaps at the level of soul, like attracts like. At the level of particle or body, complementary opposites attract.

It is *when we literally synchronize our lower personal fields with
the higher vibrational energies of the
multidimensional and quantum self
that we begin to join the more synchronous, coherent, and creative fields of consciousness. It is the
soul that has the power to truly create.* It is not the personal self.
The lower self, when out of synch, has to work hard and with
great effort to make much happen, because it is working with
the slowest, most physical mechanical and electrical forces.

As we approach the close of this chapter, the idea is perhaps dawning that who we are is much more than most of us
have ever remotely imagined.

In our wholeness, we are multidimensional and holographic beings, spheres of consciousness within even more
expansive spheres of consciousness, with the power to create that which we can coherently imagine. Within time and
space is the slower vibrational personality or personal self.
This is the self we are used to thinking of as the self. Lim-

ited to our physical body, emotions, and personal thoughts. Simultaneously, originating nonlocally and holographically beyond time and space, exists the faster vibrational, transpersonal, spiritual self.

As the physicist Amit Goswami teaches, we really are a quantum self.

In the next chapter, we will consider four of these transpersonal centers of consciousness, along with the three personal, and how they are intended to work together to bring us our sense of self, our various types of intelligence, motivation, and creative activity.

Let's proceed from our human perspective to map our seven major spheres of consciousness, the information that they bring into our awareness, and the ways they relate to the progression of our human development. They are the immediate sources for our values and motivations, our fears and desires, and our various types of intelligences. From them arise the internal conflicts most of us feel. And hidden within them are also our greatest powers and potentialities, sources for our highest inspiration and joy.

Chapter Two

Layers of the Quantum Self— Consciousness Map and Symptoms of Imbalance

In the previous chapter, we learned that at our core, we are power, we are energy vibration carrying patterns and qualities of information. We are consciousness. We are a quantum self with infinite layers to our being.

Technically, this "energy-carrying-information" that we are defining as "consciousness" is all around us, powers through us, brings us life, comprises every aspect of our being, and interconnects us with all that is, seen and unseen. It forms the multidimensional and holographic matrix of our being as a series of *concentric **spheres or orbs** of consciousness.* Picture nested Russian dolls—where there is a doll within a doll within a doll—with the smallest doll as our most physical self. We'll simply refer to them as **centers** of consciousness.

If you'd prefer, you can picture the energies that make up the self as *highways of energy.* Many, many highways of energy, intersecting with one another, layer upon layer (like a ball of energy "yarn"), and forming multidimensional spheres of energy, just as all planetary bodies are spherical, like the Earth. These highways are the *ley lines of our earthly body* and are *called* **meridians** *in Eastern medicine.* And, just like highways near our homes, these roads and highways are of varying sizes, large and small. Where major energy highways intersect, there are large vortices of energy that appear as "spinning wheels" of energy called **chakras.** The seven major chakras or vortices serve as the *chief intake and outtake valves for the multiple spheres of energy that make up the quantum self;* they are energy receivers/transmitters, and each energizes a particular area of the etheric (i.e., archetypal blueprint energy) body, guiding the creation of its final form.

In this Consciousness Map of the self, we are going to talk very briefly about the seven fundamental centers that are most relevant to the way we seem to develop our psycho-spiritual awareness and understanding over time and how we fall into imbalance; the types of information and intelligence predominant within each; and how each of these seven centers motivates our life actions, influences our most strongly held values, and therefore sheds light on why we love what or whom we love and what leads us to feel happy.

As discussed earlier, the three slower or lower vibrational centers or spheres of our ego body are a dim reflection and projection of the higher light or soul consciousness. The higher frequency spheres of our holographic and multidimensional soul serve as the archetypal blueprint (for the lower), carrying the potentiality of our soul's numerous qualities within them. The lower spheres serve as vehicles of expression for the higher, transmuting and bridging soul energy into this material plane of ego for creative purposes.

As this map becomes clearer in our mind, it will give us a sense of how we are similar to and different from others, why we like what we like, where we are in the expansion of our consciousness, where we might grow next, and why we feel such conflicting parts within ourselves that lead us to get stuck, confused, and frustrated.

It is important to keep a nonjudgmental attitude as we proceed. *None of the centers is more or less important for our development than any of the others. They each serve a vital purpose* in this physical plane of reality and are intended to be integrated together. It is when the centers are not in balance or synchronized as a whole that our problems arise. We then tend to focus blame on one or another center, especially the egoic centers, and toward those people who are predominantly living from centers that do not reflect our own values. But *ego is not your enemy. It's the only vehicle for expression in this plane that you have.* Its job is to act in service of your creative soul.

In the following illustration, note that it is impossible to really convey the complex patterned energies of the holographic quantum self. Yet allow this image to give you a sense of the nested spheres of energy, with many vertical swirls of force intersecting with the horizontal. Here, for clarity's sake, I have limited the diagram to seven horizontal rings but have tried to capture the increasingly encompassing nature of our spheres of consciousness.

The intersections of the vertical with the horizontal give you a sense of the merging electromagnetic fields that radiate outward from every atom, cell, organ, and body, just like it does with the body we call Earth. The spiraling is intended to give you a sense of the twisting motion of torsion energy, which seems to serve as a wormhole of sorts as energy is transformed higher and lower into its multiple frequency bands of expression.

As you climb the ladder of human consciousness, you'll find yourself traveling from the darker, most dense and material regions of the personal self into the lighter, more subtle energy territory of the transpersonal self. The higher, broader spheres include the lower, narrower realms of experience, which is why we can understand others in personal consciousness more than they understand us within transpersonal consciousness (though in reality each sphere is always there for each person to tune into).

Within each layer, I have noted the type of awareness or consciousness, with a few keywords of experience, as will be discussed in this chapter. The lower three spheres of localized personal consciousness seem to follow rules of time and space, and therefore are governed by electromagnetic energies. However, the higher four regions of nonlocal transpersonal collective consciousness do not obey the conventional rules of time and space and seem to involve energies that effect "spooky action at a distance," what quantum physicists are now studying as torsion fields. No wonder when we find ourselves in higher vibrational transpersonal consciousness, we experience intuitive guidance, Law of Attraction, synchronicity, telepathy, precognition, and other such "impossible" nonlocal experiences!

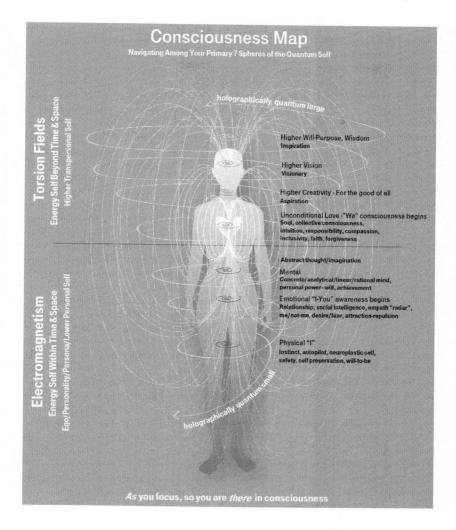

Consciousness Map
Navigating Among Your Primary 7 Spheres of the Quantum Self

holographically quantum large

Torsion Fields
Energy Self Beyond Time & Space
Higher Transpersonal Self

Electromagnetism
Energy Self Within Time & Space
Ego/Personality/Persona/Lower Personal Self

Higher Will-Purpose, Wisdom
Inspiration

Higher Vision
Visionary

Higher Creativity - For the good of all
Aspiration

Unconditional Love -"We" consciousness begins
Soul, collective consciousness,
intuition, responsibility, compassion,
inclusivity, faith, forgiveness

Abstract thought/imagination

Mental
Concrete/analytical/linear/rational mind,
personal power - will, achievement

Emotional "I-You" awareness begins
Relationship, social intelligence, empath "radar",
me/not-me, desire/fear, attraction-repulsion

Physical "I"
Instinct, autopilot, neuroplastic self,
safety, self preservation, will-to-be

holographically quantum small

As you focus, so you are there in consciousness

Center One: Physical

The mystics might say that God exhales us into this life as
an extension of his one life, pauses his breath as we play in this
domain of form, then inhales us back into the bosom of his
love and into the omnipresent awareness of the formless state.

In more quantum language, we can say that as a soul, we have condensed out of the quantum field of potentiality. Now, as a result of the soul's intent and creative purpose, we find ourselves holographically projected here from a realm of Wholeness consciousness into a quite containerized human body.

Doesn't it make sense that, from our human perspective, we have to first tune into that which is physical and material?

Center one consciousness is our **physical** consciousness. Within this circle of awareness we have the *material felt **sense** of "I,"* the physical perception of our dense body, its boundary or skin. From our human physical perspective, it is rooted through the red energies of the first chakra at the base of our etheric spine.

This is the center of the self that is *predominantly focused on attending to the physical and material needs of the individual* self. For that is how our localized physical self in limited awareness perceives itself, as *separated* from all that is and alone to fend for itself. That is what guides its internal motivations, values, and life activities.

With the perception of such separateness comes the sense to protect the physical self. So no wonder we first *value physical health, safety, security, and **self preservation.*** It is logical that we are moved to seek out food to sustain the body, to find the comfort that comes from shelter and clothing. In today's world, that means money, so of course we feel compelled to make money. For with it is our means to acquire that food, shelter, and clothing that we need to feel safe and secure.

When we resonate primarily with this center, it seems that our *power comes externally from physical* means, including physical control, aggression, dominance, competition, and the like.

Within this physical center of awareness, we pursue *happiness* by accumulating physical things, like more clothes, bigger houses, better cars, and the latest toys. We crave the physical sensations in the body that arise when we partake of delicious food and drink, get high, or purchase something new. We alter our mood biochemically if we have to. However, since material pleasures, being physical and biochemical in nature, are fleeting and impermanent, happiness stemming from this center is short-lived.

It is the same with *love*. In the physical sphere of consciousness, we seek and express love through physical sex and having children. We equate love with buying and receiving gifts and spending physical time with one another. But the love we deeply desire continually seems to elude us.

This center of consciousness is largely our **instinctual** intelligence, the *collective intelligence of our bodily cells.* Instinct usually operates subconsciously and below the threshold of our broader, more conscious, ego awareness, in order to carry out all of the energetic functions of the body—its heart; its brain; and all of the organs, muscles, tissues, and bone. In terms of neuroscience, this is the realm of our **neuroplastic self,** the process of the brain continuously wiring and rewiring its entire nervous system to accommodate what we learn and whatever it is that we do repeatedly, so that we can do it fast and efficiently. (No wonder we develop "tolerance" to anything we do repeatedly! Then need more and more of whatever it is.) This is the **autopilot self,** the operating system that comes preinstalled in this body and is in command until it is superseded by our higher intelligence.

Once the necessary physical consciousness is anchored, we are ready to perceive our more psychological nature, that which comes from our emotional and mental energies. Together, the physical with the emotional and mental, these three centers form the sphere of consciousness that Western psychology refers to as the personal self, the ego, persona, and the personality.

Technically, our human development is not this linear. All spheres of consciousness are always there for our perception and awareness, but most of us aren't aware of them all. Most of us move in perception more from our physical, then into emotional awareness, so that is the order in which we will look at them here.

Keywords of Center One: Physical, "I" awareness, instinctual/cellular intelligence, autopilot self, neuroplastic self, the color red. Primary values and motivations are for physical pleasures and physical health, safety, security, and preservation of the physical body and personal self (e.g., food, shelter, clothing, money, sex).

Common Symptoms of Imbalance in Center One: Unconstructive habits and addictions, not being able to support oneself, reduced trust in self, insufficient independence, undeveloped skills and abilities, too little/too much physical activity, laziness, physical aggression, overemphasis on competition for material resources, imbalance in sex drive, suicidal behaviors, overly risky behaviors, impulsivity, excessive fear of physical danger, material greed, material control/dominance over others, overall poor health, illness due to difficulty handling excessive toxins in the world, excessive sense of feeling separate/limited/alone in the world, isolating or pushing out the rest of the world.

Center Two: Emotional

Center two awareness brings with it a sense of *"I-You" consciousness.* Ready to perceive more than just the "I," we begin to notice the people and things around us. We begin to notice that we are in **relationship** with all that is around us. This is the center from which first arises our *value for relationship, family, friends, and belonging.*

As we *focus our attention on the relationship between us and some external* person, or object, or idea, or color, or sound— anything at all (e.g., our significant other, a pattern on a new sofa, ideas about politics, the colors red or green, classical music)—we will *notice a movement in our body* of some sort. It is these particular movements, or **merging energy vibrations,** that we label with specific *names of emotion.*

That is, the energetic vibration of "I" merges with the vibration of "you" or "other," and results in a blended field of energy. "I" will feel some degree of resonance as a result. This is how we feel varying *degrees of me or not me,* attraction or repulsion, like or dislike, desire or fear. That which is most similar to our energy will feel good to us, because it resonates with our inner essence. That which is less similar will, to some degree, feel neutral or unpleasant.

Empathy arises from energetic resonance and entrainment with another. When we are open to perceiving emotion within ourselves, we can not only feel our own emotion, but we can also begin *to sense the emotional energy emanating from others* as well; it is the *early sense of psychic* experience. As we allow ourselves to resonate with another, energetic information is exchanged among us, and we have entrained to the other's field. It is like we are one, and our energies have become one collective field. (When we walk into a crowded room, and we are perceivers of emotional energy, we will at once have a sense of the collective energy in the room.)

It is no surprise that those who are sensitive or gifted at feeling emotion will have more of the sense of **social intelligence**. (When they are balanced in the centers. When they have not yet learned to balance the centers, then they likely will numb or dissociate from the emotional energy that feels overwhelming to them. Or they can appear in the extreme to be emotionally volatile, extremely moody, or even unstable.)

As we feel emotional energy, it moves us, literally. It is this energy that *motivates* us, drives us, impels us *toward* that which resonates with our soul and that which we like or desire and *away* from that which we dislike or fear. From this center, we are swept into groups of similar-minded others and away from those who feel dissimilar or even threatening.

Within this center of consciousness, power seemingly comes from our external relationships. Others likely see us as charming. We may tend to be extroverted and get energy from other people.

Happiness and love stem from being with our family and friends. We yearn to have children of our own. We love the emotion of feeling "in love," and being swept away by resonance with another person.

Mystics say that center two is anchored through the chakra system in the etheric area of our reproductive organs and is orange in color.

I would say that most of humanity in today's age is predominantly living from centers one and two. That is why there is so much emphasis around us on physical and material comforts, sensory experiences, and socializing to find personal relationships from which to form families and to have children. That is also why there is so much emotional and instinctual reactivity, impulsivity, aggression, physical competition, and addiction in all its various forms. We get stuck in centers one and two and hold ourselves back from more advanced ways of being and interacting with others.

Keywords of Center Two: Emotional health, "I-you" awareness, relationship, the lower heart, social intelligence, empathy/lower psychic energy, desire/fear, me/not me, attraction/repulsion, family, procreation, color orange. Primary values and motivations are to feel surges of emotion and to be in relationship with others. We seek connection through relationships and belonging to groups of others.

Common Symptoms of Imbalance in Center Two: Dependency/codependency/enmeshment in relationships, patterns of relationship addiction or other issues, emotional instability or volatility, reduced motivation or emotional drive, excessive fears or desires, fear of being alone, social anxiety, problems with intimacy or sex, inability to feel empathy or emotional energy, emotional numbing/dissociation/avoidance, emotional eating, attachment issues, excessive pleasing behaviors in relationships or trying too hard to fit in with others, unwillingness to stand out as unique or different, diffi-culty trusting others, promiscuity, and addiction to lust, sex, romantic "love" or being "in love," conditional and selfish "love."

Center Three: Mental

This is the sphere of consciousness from which springs our **concrete intellectual** ability to *think linearly, to rationalize, and to analyze*. It makes sense that we need to think linearly, because on this plane of consciousness, we are in time and space, and most of us aren't telepathic. We are forced to com-municate in linear words and sentences. We are forced to use our "left brain" and to solve problems with this layer or center of our intelligence.

From this center, we are able to take ideas that float into our "right brain" mind whole via higher vibrational abstract thought, and to separate them into individual thoughts. This is lower vibrational mental consciousness. It is *mental energy reduced (i.e., transformed, transduced) and chunked into **indi-vidual thoughts.***

It is concrete thought. *Mental energy **pointed at the con-crete physical** world and concerning the personal self.* This part of us is usually cultivated from early childhood. We learn to think before we act. We discover that our actions have consequences. We learn skills so that we can apply them in increasingly creative ways, preparing us to move past simply supplying ourselves with food, shelter, and clothing. In this way, we come to know that we are truly powerful and that we can create something of our choosing.

As we grow into mature adults, we expand our minds into

more complex streams of thought energy, logic, dedu... reasoning, abilities to compare and contrast, analysis, an... problem solving in our everyday lives.

All of these are essential intellectual life skills necessary to interact with and create within our physical world.

Yet also from here springs our *"monkey mind,"* as it is called in Buddhism, because thoughts for most of us seem to jump around from one thing to another, from topic to topic and idea to idea, like a monkey races from branch to branch. Most of us have not yet learned to discipline our thinking. To try to stop certain thoughts or to direct thoughts in a certain way seems almost impossible at first—the average human attention span is eight seconds—until we begin to learn how to concentrate and eventually to meditate. Then we can attend to what we want for as long as we choose. And as we do so, abstract thought and higher mental energy begins to grow in our awareness.

It is within this center that we develop our mental power, come to *know who we are in this physical reality,* learn to focus our **personal will,** and are motivated toward **achievement.** It is this energy that moves us to get a job, learn a trade, or find a career where we can use and expand our personal skills and abilities. It is intended that we *know the unique set of energetic or consciousness qualities inherent* within us, so that we can put them to use for the betterment of the planet. When we are in this center or state of consciousness, we will be most happy when we can see our personal achievements, for this is how we gain our sense of personal power.

Within this center of consciousness, love energy is focused toward learning, self-growth, personal achievement, and the

on or career. We have to build the ego

transcend living from the ego.

balance, the result of all of this activity

ealthy sense of **confidence, self-esteem,**

. This paves the way for the higher trans-

personal self to make use of the lower personal self, for the soul to spread its creative love energy into broader and more inclusive circles through the egoic self.

Mystics say that our lower mental energy (considered to be the synthesis of our instinct and concrete mind together) is related particularly to our etheric solar plexus, the energetic area around our stomach and the enteric nervous system. Scientists have confirmed that the enteric nervous system is its own entity and call it the "gut brain."

I can say that most of my clients who have a habit of negative thinking, of fearful thought energy causing a problem in their lives, tend to have stomach, digestive, and elimination problems like diarrhea, constipation, more frequent urination, or worse gastrointestinal types of ailments and eating disorders. So the phrase "gut instinct" seems right-on for this center. It's a mixture of first center instinct, second center fear-based reactivity, and third center thinking focused upon the lower personal self and its slower vibrational or more material needs.

The three personal centers interact in interesting ways. Together they form the basis for our broader psychological health and more **integrated personality.**

Center three brings a balance to center two and to center one. We are intended for these to integrate into a healthy, well-functioning personality. It is important to develop our mental intelligence and personal power and to learn what it is we can achieve when we put our mind to it and work through all three centers with a sense of proportion. Otherwise, we can get stuck in our most unthinking, impulsively reacting, primitive self (overemphasis on center one). Or we may become lost in the ocean of our relationships (overemphasis on center two). We can become dependent on others. We may even forget who we really are, a unique fingerprint of energies carrying into our being and this world a certain blend of qualities and characteristics.

Once we have learned what we can do with our personal power, it is then that we are ready for the developmental task of the soul. For the soul, as we will see in center four, seeks to take what it has developed through its personality (the combined energies of the lower three centers) and use it in service of the larger group we call humanity and for all of the Earth's citizens.

Keywords of Center Three: Mental health, lower vibrational mind or mental energy, concrete/rational/linear/analytical intelligence, concentration, elementary abstract intelligence, confidence, self-esteem, personal power, persistence, meditation, determination, initiative, personal will, leadership, achievement, the color yellow. Primary values and motivations are personal power and achievement, desire for learning, and gathering data/knowledge.

Common Symptoms of Imbalance in Center Three: Aggressive career competition, insufficient self-assertiveness, imbalance in personal power, addiction to achievement, overidentification with the thinking self, overreliance on "left brain" thinking, monkey mind, inability to sustain mental focus, learning disorders, thinking in extremes, overanalyzing, ruminating, difficulty using imagination or thinking "out-of-the-box," avoidance or addiction to learning/knowledge/data, insufficient confidence or self-esteem, unconstructive self-talk and negative self-evaluation, criticism, contempt, disrespect for others' ways, bragging/boasting, arguing, having to be "right," stomach/digestive problems, comfort eating and food addictions (trying to fill the "emptiness" of the third vibrational personal power center anchoring in the area of the gut).

Center Four: Unconditional Love

This is the center of **unconditional love**, **"We" aware-ness,** group mind, collective consciousness, **intuitive** intelligence, and **soul.** Within this state of love consciousness, we are in energetic resonance and nonlocal entrainment to spheres of consciousness well beyond the personal and certainly beyond the physical boundaries of time or space.

As soul, tapped into group mind, united within the energy of love, we are as if **telepathic** with all of those of similar vibration, no matter how near or far they are physically and materially from us. We are extremely sensitive in our conscious awareness to what is not only within our personal mind but also within the greater sphere of mind that now encompasses our sense of being. In this state, we are trans-personally, holographically aware of ourselves, both as unique individuals and as a part of a larger group.

As we initially awaken to this nonlocal soul consciousness, a sense of **responsibility** to the group floods us. Over time, we more readily take on the responsibility to do what we can do *for the good of the whole.* We *aim to serve* and *value all others equally* in their service.

This is the sphere where **unconditional love for all without exception** *is the motivating impulse for life.* From here we are inspired to love for love's sake, to love without consideration of personal gain, to love even when hurt, rejected, ridiculed, abandoned, or forsaken entirely. Within this state, we hold steadfast to love as our way of engaging with the world. We strive to be gentle, kind, authentic, honest, patient, generous, and respectful of others.

From this sphere of consciousness, we crave to get along with others and to find common ground as we go about our work of helping. We seek to **cooperate** and **collaborate** with one another. Our love is allowing and freeing, instead of controlling or manipulative as it is in our personal consciousness. We appreciate the way that *collectively we achieve perfection through the bounty of the diversity* contained within our one Body.

We focus on **sharing** the gifts we know we have and on welcoming the gifts of others as we go about our collective or group work of improving the state of the planet in which we live and have our beingness. We appreciate the way major advances take place one small step at a time by each and every person, no matter the size of any particular role on any given project.

From this center, **compassion** is natural. We know we've made about every mistake we could make over the course of our lifetimes in various dimensional realities of limited awareness and that any mistake we or others make is just a part of our learning and a matter of experience. And we know that the suffering that is experienced in this world serves as a vehicle for learning personal and collective compassion and forgiveness.

Forgiveness is common sense in this state. For why would we hold ourselves or others hostage to any past mistakes when instead we could learn from them and more simply move on?

Moved by higher love, *we naturally let go of personal needs and desires, for they simply don't have the pull they once had upon us* when we were limited in identity and awareness to the physical, emotional, and mental layers of the personality consciousness.

Our sense of power shifts. Instead of our *power* predominantly coming from the personal mind in center three, as well as the other two personal centers of activity, it now comes *from the* **coherent power of unconditional love energy** and the nonlocal field of *"We" consciousness* inherent within this center.

When we are in present-moment awareness, it is easier to access this center. The lower mind stills and becomes quiet and receptive. The inner ear is poised, listening for that still voice inside. The loving heart is open to be moved by the subtle energy that we detect from here. We begin to intuitively sense the direction in which we are being led.

From this point in consciousness, we practice **faith** in the more encompassing consciousness that is the greater symphony in which we are playing a part. And it is from a loving and trusting heart that we come to *intuitively* play our notes as we hear or feel them. We come to trust that the conductor knows the purpose for our body instrument and the timing of our participation.

It is through this door that most of us first step into **transpersonal awareness.** It is this particular plane of consciousness that is religiously symbolized for our concrete brains by the sacred ritual of baptism. It is this transpersonal state of mind that makes us *feel "born again."* Within this place of soul consciousness, we feel as if we have been born again, for we indeed have been moved, quite literally, into an entirely new way of perceiving, loving, and living in this world. When we have actually experienced this state, we don't feel the necessity to talk about it, because we are urged instead to get to the business of acting upon it. We feel a responsibility to bring love into the world, steadfastly, constructively, and unconditionally, including all others within our sphere of love. When we are truly within this sacred spiritual state, there is no one who is excluded from our love.

Within this sphere of consciousness, **gratitude** is easy. We feel *interconnected with loving presence,* and *sense our part in the whole.* We have a budding sense of meaning in what we do and see the synchronicities unfolding around us and guiding us toward our higher purpose. As we take all of this in, we bask in appreciation for All That Is.

Mystics see the energy of center four as radiating emerald green throughout the etheric heart space.

Keywords of Center Four: Unconditional love, "We" awareness, group mind, collective consciousness, soul, transpersonal consciousness, higher heart, intuitional intelligence, faith, love for the sake of loving, wise love or Christ consciousness, compassionate understanding or Buddhic consciousness, sense of responsibility, ability to forgive, letting go of the personal self, gratitude, appreciation, respect for diversity, budding motivation of cooperation/collaboration, color green. Primary values and motivations are to practice unconditional love toward self and others, to maintain and radiate the higher energy states of love and gratitude, and to take on responsibilities that purposely serve the collective common good.

Common Symptoms of Imbalance in Center Four: Difficulty feeling a sense of love or connectedness within oneself, inability to love beyond immediate circle of family/friends, avoidance of cooperation/collaboration, inability to sense intuition, confusing instinct (fear-based gut reactivity) for intuition, lack of compassion, avoidance of responsibility, insufficient giving of oneself, difficulty in receiving, insufficient self-love, unconstructive loss of self, unwillingness to compromise, struggle to forgive self/others, inability to offer constructive and loving feedback rather than criticism or punishment, disrespect for diversity of life, self-centeredness or selfishness/narcissism, feeling overwhelmed when trying to love unconditionally at the highest levels, "perfection" thinking, heart/lung issues, Dark Night of the Soul, dark night of the senses, spiritual or transpersonal depression, despair at the lack of love in the world, spiritual or transpersonal anxiety.

Center Five: Higher Creativity

This is the center of **higher creativity.** As we awaken into the transpersonal consciousness of center five, our sense of creativity shifts. Instead of creating from the lower vibrational and personal centers, which are motivated by individual power, gain, achievement, and the urge to procreate, we now exercise our power by *creating for the good of all*. We are motivated to help and serve the world at large in our own unique and soulful way. And we are inspired to work together in creative loving cooperation and collaboration with others who hold similar visions and values.

By creativity, I don't mean the word to be limited to painting on canvas, sculpting stone, writing books, or even designing great buildings. I mean it to be *any loving activity that is constructive toward evolving ourselves, humanity at large, or even the planet. These are our higher acts of creativity.* In reality, the majority of my clients who have touched upon their soul consciousness in center four are still seeking to find their fifth center calling and creative purpose. It is because they have been swept toward the consciousness of center five but have yet to anchor it into their daily awareness. They have yet to be consciously aware of their **core qualities** *and inspirations as a soul.* As we anchor center five energies into our conscious awareness, through contemplation and meditation, we inevitably find out who we are as a soul and what it is we are called to create.

When we live from center five, *we allow the intuitive loving intelligence,* our collective consciousness or "knowing" that we developed in center four, *to guide the* **actions** *of our daily lives.* The *higher heart is now our CEO,* which is in touch with the "higher-ups" who know the importance of a particular direction; and *the head (aka linear mind) serves as our administrative assistant,* helping us carry out the linear day-to-day tasks necessary to make our ideas reality. For example, the heart may urge us to study a particular major in college, and the head helps us work out the details and do the paperwork.

As a soul, it *feels imperative that we follow our* **calling**. It is our new priority. We have to do what our higher heart begs us to do, for it knows the purpose and reason we have been born.

Creating at the highest levels of our being is like nothing else. In this place of consciousness, we are *starting to feel more aligned* than ever before, with at least five spheres of conscious awareness and distinct intelligences at our fingertips. We feel passion moving through us, we sense our purpose and reason for existing unfolding before us, and there is ever so much more meaning in our life.

Mystics say this sphere of consciousness is a sapphire blue and associated with the area of our etheric throat. From it, we really start to find our true and inner voice.

> **Keywords of Center Five:** Creative intelligence, higher creativity or creativity done for the good or advancement of all, aspiration, cooperation, collaboration, color blue, trust in and movement from that which is higher, budding sense of "thy will be done." Primary values and motivations are to live an inspired and soulful life, create collaboratively with others, and to improve the world in some fashion.

> **Common Symptoms of Imbalance in Center Five:** Spiritual ambition/overachievement/frustration, inability to be creative, unwillingness to stand out as unique in perspective or creative ability, too little risk-taking to act on intuitive guidance and creative impulses, feeling of leading a less than soulful life, throat/thyroid issues.

Center Six: Visionary

The sphere of consciousness that we are calling center six is said to be connected through the chakra system in the etheric area between the eyebrows, a point where the blueprint energies architecting our two physical eyes are integrated as one vortex or center of energy. It is said to be the indigo color of the midnight sky and is referred to by mystics as the **Third**

Eye. It is referenced in the Bible by phrases such as "when thine eye is single."

This "eye made singular" is evidently a *state of laser-like and coherent focus of consciousness*. One that takes us on a *quantum leap into abstract intelligence, imagination, and the world of the unknown*. One that gives us the power of ***higher vision***, *of being able to see what has yet to be concretized or manifested into our physical reality*.

Those who are gifted in this sixth center energy are typically labeled as *strategic planners, visionaries, inventors, prophets, and geniuses*. Between the time of Copernicus and Galileo lived Giordano Bruno, a Dominican friar of sixteenth-century Italy, who taught from and died for his Third Eye vision. As history tells it, he experienced a vision in which he saw the cosmos as comprising an infinite number of universes, each filled with a multitude of intelligent beings. It is possible that he may have been one of the first to describe the multiverse as it is theorized today—the infinite array of life that is said by quantum-edge physicists to be our true cosmic reality.

All of history's most renowned scientists and creative geniuses may in fact be moved by the tremendous power of this center of **visionary consciousness.** Tesla and Einstein. Tchaikovsky and Mozart. Just to name of few of the obvious. When the Harvard sociologist Pitirim Sorokin wrote about such creative geniuses, he concluded that what they each seemed to convey when interviewed was the notion that the *ideas seemed to just come to them, to occur to them, to pop into their head.* They described that the music or vision or solution to the problem they were attempting to solve merely dropped into their mind or came to them in a dream, and they simply wrote it down.

Tchaikovsky added that there are inevitably places where we have to fill in some missing pieces of what we have received intuitively; when we do so with our lower mind, it is noticeably inferior material, but it is simply part of the process. (How many Third Eye visions have you received and promptly forgotten or even dismissed?)

As we tap into the unfathomable reservoir that is consciousness as a whole, we owe it to ourselves and the planet to bring it back and share it for our collective evolution and enlightenment.

Keywords of Center Six: Laser-like and coherent Third Eye, single-mindedness, visionary, seeing what is not yet manifested, higher psychic, indigo color. Primary values and motivations are to pioneer new frontiers, to take mankind where it has not yet been, to move the world forward in some way.

Common Symptoms of Imbalance in Center Six: Insufficient focus or persistence in seeing one's creativity through to completion, inability/unwillingness to conceive the intangible, too little abstract intelligence or imagination, limiting oneself to linear/logical/deductive reasoning processes, problems balancing or getting lost in the psychic senses (including the detection of paranormal phenomena).

Center Seven: Inspirational and Transcendent

This seventh center is said to stream through the etheric crown chakra above our head and to be predominantly violet in color. From this center and through the gateway of the seventh chakra, we are *nonlocally and holographically interconnected with all fields* of consciousness, all senses of intelligence and life energy, as they cohere the archetypal blueprint for our energetic or etheric heart, brain, and body.

Mystics tell us that it is *through this portal that our soul consciousness leaves the confines of the body when it is asleep, for the soul has no need for sleep. And through here that we have near-death experiences (NDEs), out-of-body experiences (OBEs), peak spiritual experiences, and other such transcendent states and visions.*

In this center, we are in a state of **transcendent consciousness,** when the personal sense of self seems to dissolve in awareness as it is superseded by the focus in higher vibrational and more encompassing spheres of collective consciousness or beingness.

These experiences can be as distinctive as we are and as diverse as the many spheres of consciousness to which we have gained access. Sometimes through this center, we *become One with All That Is*, variously described as an *alignment* with God Consciousness, with Presence, with the "I AM," with Source, or with the *quantum vacuum or field of pure potentiality.* We may experience *other dimensional* worlds, life on different planets and in uncharted galaxies or universes. We may visit bandwidths of pure energy, sound, light, color, or geometric forms. We can have the experience of being an energy or light body—even a star or sun. Or we may enter the nothingness of the "Void," a state beyond any physical perception or transcription.

Yet, too, from center seven, we may find ourselves as the consciousness of an atom, a vascular system, a rock, a mineral, a plant, an animal, or some other form of life other than our present one. We may experience our own consciousness in the womb or in a different time of our current life, maybe past or future. We might even enter the experience of another lifetime, either our own or the imprint of someone else's. *As we point the focus of our consciousness, so we are **there**.*

From this center, we feel **inspired** because it is literally so that we have breathed in Spirit. A higher sense of **wisdom consciousness** and **purpose** comes from here. We feel a downpour of *synthesis*, of *knowledge that is wise and full of loving integral understanding. Divine Will becomes our will* simply because our will is synchronized with such higher states.

Others will say we live a life of **sacrifice.** But to us it is not a sacrifice. It is our joy. Maybe we give up having a family and raising children, because we are serving our global family. Maybe we offer our lives to attend to the sick, as did Mother Teresa. Maybe we lose our life for our mission, as did Friar Bruno, Abraham Lincoln, Martin Luther King, and John F. Kennedy. Maybe we spend our life in prison for our beliefs and cause, as did Nelson Mandela. Maybe we dedicate our life to human rights, as did Mahatma Gandhi, and fourteen-year-old Malala Yousafzai. In this state of consciousness, it does not feel like sacrifice because we are *not attached* to such things as material objects, personal achievement, or even the physical body.

In this state of transpersonal consciousness, we feel *free*. We are free, because we *know* from experience we are infinite. We are free, because we are no longer constricted to the limited beliefs and set of values of the personal self. We feel free to act upon and live from our transpersonal consciousness, free to be who we are regardless of the lack of understanding from others.

Keywords of Center Seven: Portal for consciousness and life energy, color violet, center of inspiration, portal for out-of-body experiences (OBEs) or near-death experiences (NDEs), transcendent consciousness, alignment with God/Presence/the "I AM," oneness, the sense of interconnectedness with All That Is, the higher sense of wisdom consciousness, the budding sense of Divine purpose and Universal will and the value of "thy will be done." Primary value and motivation is for Divine Will to work through all of the seven centers (uncontaminated and unfiltered) in order that Divine Purpose may be manifested through us.

Common Symptoms of Imbalance in Center Seven: Feeling uninspired, imbalanced state of willpower, unconstructive use of substances to get "high" or to feel the connectedness of the higher states, overattachment to transcendent experiences, overattachment to OBEs/NDEs, dissociation from body especially in cases of severe trauma, loss of groundedness leading to apparent psychotic symptoms, loss of ability to navigate between transcendent and material realms, spiritual addiction, comas or insufficient bodily consciousness.

Chapter Three

Nine Varieties of Transpersonal/Spiritual Anxiety

Quantum Physics to the Rescue

Thank God for quantum physics. I'm serious.

For those of us who have been experiencing things that are supposedly "impossible" to experience, we finally have some vocabulary we can borrow when it comes to talking about how to understand and cope with such experiences.

As it turns out, we're not crazy—we're quantum!

But let me be the first to say that no matter how I write about these things, there will inevitably be scientists out there, even physicists, who will discount what I say. I'm okay with that. They are still arguing among themselves about what is hypothesis, what is coherent theory, and who is more right about our worldview.

According to Bray, "there is no solid finalized hypothesis

in Quantum Theory that is universally agreed upon." And he emphatically points out that "most of the modern interpretations of Quantum Physics cannot explain" essentially all of the data and outcomes from the most important experiments. Even the fathers of consciousness research agree that the current models of "reality," especially those of materialistic Western medicine and psychology, are not consistent with what is actually found in their experiments concerning transpersonal states.

What does appear clear is that material reality, spacetime itself, is shaped by consciousness. And somehow consciousness works both forward and backward in time. Even more oddly, the more sustained our attention, the slower that time moves at all.

Yet we still don't have a consensus definition in the sciences for either consciousness or thought, even though consciousness is evidently responsible for materializing the somethings out of the "nothing," the empty space that is anything but, and that holds more power in a fifth of a teaspoon than in all of the known visible universe.

Let me also point out that as our perception of time passes, all consensus on scientific fact is eventually superseded by new scientific realities. For instance, it wasn't that long ago that scientific consensus believed our Milky Way galaxy *was* the universe. Then Edwin Hubble, in 1922, proved that Andromeda was too far away to be in our galaxy. Overnight, scientific consensus was made obsolete. A fresh new view of our universe emerged, one filled with innumerable galaxies.

The same is happening again today, and scientists are still

trying to figure things out. Even a concept like gravity, which most of us thought was a known thing, suddenly isn't.

Enter a new worldview, a universe comprised of multi-verses and parallel realities, where physicists now sense that not even two sides of the same atom or neutron are in the same spacetime domain! Where time cannot be separated from space. And, too, where space and time seem to be illusions of perception and focuses of consciousness.

In the meantime, we're living the quantum reality of this nonlocal, multidimensional, and holographic sense of self and need some vocabulary to talk about what is going on for us and mechanisms with which to cope. So let's get to it!

Side Effects of Wholeness

The Privilege

To sit, not saying much,
giving all your attention
to the one who is in labor
giving birth to a new self,
which not long before
she did not know to expect.

To sit, not saying much,
with the tears that come,
with a greater fear and
the sweat of this struggle
not to run,
so to find the courage
she did not know would come.

To sit, not saying much,
during the seismic shaking
brought about by finally hearing
the sound of her own voice.
—John Harder, 2010

It is indeed a privilege to sit with someone who is finally coming into a new sense of self, a quantum self. Where, like a rose, consciousness has unfolded its many petals outward in a broader outreach of experience and expanded beingness from the personal and into the transpersonal.

When personal awareness unfolds or merges into the unifying field of quantum consciousness as a whole, it is generally perceived by us as a felt sense of Wholeness. We feel deliciously full, complete, and whole. Our illusion of separateness is replaced by a knowing of interconnection with All That Is. This sense of Wholeness feels like we have been embraced by God and are cradled in the arms of Divine Love.

One of my clients, a young woman in her twenties, was doubting her ability as an energy worker and was eager to do a somatic (or body mindfulness) meditation to see what she might learn about herself. By the time our session was over, she had experienced herself beyond time, where she was pure light and love at her soul Source.

While she was in that state, the energy in the room expanded such that—I exaggerate not—I had to attend to my own energy field, because I started to feel literally crushed by the magnitude of her radiation. I went within and strengthened my own light, seeing myself as a sun blazing forth unconditional love. I also gave her a suggestion that she could allow her energy to flow outside through the windows, that she did not have to contain it within the four walls of our small session room. Without speaking, she immediately followed the suggestion, and the energy of the room subsided enough to be more comfortable to me, sitting in her presence, while she was in her expanded state.

When we processed her experience immediately afterward, I could tell it was completely unexpected, yet it was an incredibly healing one for her to have. She described how it renewed her faith in God, the Universe . . . and in her own abilities to work with energy.

Thinking it might reinforce her internal experience, I let her know how her energy had impacted me while she was in that state and that I had never (before or since) been around such intense tangible energy as hers had been when she was in her Source state.

Apparently in disbelief, she sat rather dazed among the multicolored silk pillows nestled upon the rattan sofa in my office—clearly still within her ecstatic state of consciousness—smiling, contemplating, and looking at me with tears of joy streaming down her cheeks.

This was a case where being able to connect with her Source state of consciousness transformed who she thought she was. And being able to discuss the physics of how it might be possible to access such states helped her left brain be satisfied enough to not get in the way of living with this new state of identity, that of herself as infinite in nature.

On the other hand, another client with whom I had only met a couple of times told me what happened to her after her previous session, where I had introduced to her the Consciousness Map of the seven centers. She said she had been thinking of all the centers and how much she could relate to each of them, when spontaneously she was thrust into a peak moment. Because she was at work, she tried to suppress it, and that eventually worked. But while she was in that transpersonal state, she said it was so overwhelming that tears came to her eyes. She then grabbed a sticky note to hurriedly write down what it had felt like, because she knew she would want to recall it later.

She recognized that she had had a peak moment one time early on in her life; then, too, she had rushed to suppress it because it was so overwhelming to her that it caused her anxiety. She hadn't known what would happen if she let it overtake her body.

She also told me that after her ecstatic experience, she came to realize that, within that state, she had come into her true and authentic self. She now wanted to live life from that state.

But she also felt anxious about what it would do to her life. "Would her family and friends reject her?" she wondered. She revealed how she now could see that she had lived a persona her entire life; she had put on a mask and created a self that others could relate to because she had always felt so different . . . and alone . . . that she didn't know what else to do at the time. Now, however, with support, she felt inspired to live her authentic self and deal with the consequences, but, she added, "I need help coping."

As it turned out, she needed help coping with the fear that she wouldn't fit in with her current social group and would lose all of her closest friends, the negative self-talk or self-judging of her ego self, the perfectionistic expectations she had of her new self, the idealized love she was waiting for, the shift in values she now felt moving her into a new line of work, the fear of failure and rejection she felt at the very thought of switching careers, the inner sense of duality and conflict she felt, the attachment she now felt to her transpersonal nature, the loss of groundedness she could sense when her energies were all cohered in her upper body, and many other issues. In essence, she needed help balancing the higher self with the lower self and the myriad emotional vibrations that played within her body.

Just as with this client, emerging suddenly into this quantum self—this higher self, this living wholeness experience of the nonlocal and holographic self—can have its side effects, particularly for those of us who come into this awareness through direct personal (or rather, transpersonal) experience and not intellectually through reading others' accounts, and for those of us who are adults when it happens.

Transpersonal or spiritual anxiety and depression are two of the most common problems, taking many twists and turns on those themes. In this and the following sections, we address these and other issues that stem from having experienced Wholeness.

Nine Varieties of Transpersonal or Spiritual Anxiety

For the body, transpersonal or spiritual anxiety feels like any other deeply penetrating fear or anxiety. But for those of us going through it, it is different in flavor and cause. Here are some of the concerns I see the most.

1) Reality Has Been Turned Upside Down

You've had a belief-shattering transpersonal or spiritual experience. The sort that contradicts what you once thought was true about the nature of reality.

You may or may not have ever felt religious before your peak experience. If you have not, you may now relate more to the term *transpersonal* to describe your shift in consciousness. If you have felt religious before, you probably refer to your "awakening" as a *spiritual* experience.

The very fabric of your thinking may be shredded. You begin to doubt everything you thought you once knew. Your perceptions of the world appear to be thrown about in a chaotic state of ruin and reorganization.

Your entire sense of identity may even crumble, the foundation of your former self gone. This is what has been referred to by many as death of the ego, and it can throw the experiencer into a state of panic and confusion.

This is a situation where you had been without conscious awareness of the higher centers, and were living through centers one, two, and three. Then suddenly centers four and beyond came bursting into your blown-open mind, bringing with them all of their information, values, motivations—all of the love! You thought you were separate, alone, and in need of competing with and protecting yourself from others. Now you realize that you are they, and they are you. Up seems down, and down seems up.

You used to prioritize material security, financial wealth and abundance—a center one motivation. Now that seems selfishly hoarding, and you might be freaking out about it, moved to change yet afraid to.

Your fourth center heart sees the entire world as your immediate family and wants to help those less skilled or fortunate, but your relatives call you eccentric and may even seek legal or medical means to keep you from giving away your wealth.

At the level of the second center, the people you used to hang out with now seem shallow and superficial, but you don't want to judge them. You just want to find a new group of folks to socialize with, people who get you and are on your same wavelength, those who have the same aspirations as you do and who prioritize the same kinds of values.

Intellectually, in center three, it now bores you to talk about other people, celebrity gossip, or reality TV drama, especially that which is mean-spirited, aggressive, or even quite primitive behavior. You turn off the news, no longer able to stand its focus on negativity, as it really does bring you down in vibration. Instead, you crave to have meaningful, more uplifting conversations focused on purposeful helping activities. You long to use your intellect to solve big problems, find solutions for humanity's most immediate crises, and heal the Earth's physical and atmospheric environment.

Your fifth center calls you to be creative in this broader way, rather than to focus on building up yourself or your nuclear family to the exclusion of your global or even cosmic family. At the same time, it is scary, and you think "who am I to move forward in such a big way?" Your ego gets in the way, its judgmental and limiting self-talk is a nuisance.

Yet the ideas and visions continue to pour in, energized by the strengthening sixth center. The more you act, the more they fill your mind. When you neglect to act (typically the result of third center negative self-talk and second center fear), the more you squirm with a sense of failure . . . and the anxiety mounts. Your creative juices then seem to trickle down or even dry up altogether.

The more you meditate on this wholeness, this unity consciousness, it is increasingly easy for you to get there. Your seventh center is a whirling generator of power. You may feel called to take classes in energy healing to better understand this energy that you wield and to share it in some useful manner. You may even get addicted to your meditations. Yet if you stop, paralyzed by second center fear of not fitting in with others, then it seems to leave you. You grieve its apparent departure, though it is really an illusion of your senses and nothing more than a departure in the focus of your consciousness and intent.

Initial Actions/Tools for Coping

When you feel acutely shaken in your sense of reality, it is time to ground yourself, especially in center one, your physical body. Yet respect all of your centers and their pulls for your attention. Take turns meeting their needs.

- There are many ways to ground yourself physically. Go outside and plant your bare feet on the moist grass or soil for a few minutes. This allows your electromagnetic field to be cleansed (you'll take in free electrons that serve as antioxidants in your system) and to reboot itself by synching up with the Earth's field. Get out in nature and unplug from technology as needed to rebalance your auric field. Or, while either standing up or sitting down, imagine roots growing out of the soles of your feet and into the core center of the Earth.

You can also bring mindful awareness to your core (for most people in the center of the body at the level of the naval, center three, but for others at the level of the heart), which feels stabilizing and centering. Get plenty of fresh water and nutritious foods.

- While you are most anxious, always attend to your most basic needs until you're ready for more. With practice, you will find it easier to allow your lower centers to be in service of the higher. For now, you may need to focus on getting enough sleep, fresh food, water, sunlight, and exercise.

- When you are ready and well grounded, then it is time to reflect on the new life you may feel called to actualize. Start by noticing which unconstructive and limiting self-talk is getting in your way and work to bring your thoughts into more supportive and "building" directions.

- And remember, each of us evolves gradually, step by step, so don't get too far ahead of yourself, or you'll just shut down from being overwhelmed.

2) Inner Conflict and Duality

Now that you've recently connected with Soul and Source, you feel a turbulent conflict within. One that knots your stomach with every twist and turn, until the highest heart brings the peace that calms.

Your faster vibrational layers propel you in a new life direction, as described above, but the stubborn ego wants to continue with what it has been doing all along, with its autopilot ways or neuroplastic normal habits. Quite often, you feel both states, higher transpersonal and lower personal, all at once; very aware of your personal desires, centers one through three, but also painfully aware of the call to sacrifice your individual wants for the larger good, centers four through seven.

This is a state of duality consciousness that has been called "standing on the razor's edge." It is indeed a thin line between ego and soul for you at this stage of your awakening.

You sense the part of you that aspires for higher, more meaningful or spiritual values. Life actions that benefit the greater good, that are loving and more altruistic. But you are also painfully aware that a part of you still yearns for wild sex, drug highs, lots of money, the kind of success that makes you feel better than others, and a lot more you would never admit to your spiritual friends.

If you consider yourself religious, you may even feel ashamed each time you catch yourself thinking "unclean" thoughts. You beat yourself up, punishing yourself with harsh and critical self-talk, believing it to be the only way to become a "better" person. Yet all it is doing is keeping you down, holding you back from the very loving state you would generally prefer to be in.

More typically, you feel many states of the spectrum of consciousness arising and subsiding, as if you are many different people arguing for control. You're in one state as you awaken in the morning. You go to work, and you're in another. You go to church to feel uplifted, only to fall as you leave the parking lot, cussing at the guy who cut you off. You still have center two moments when you feel darkly anxious, concerned, worried, sometimes even panicky. You're irritable, frustrated, and angry. Sadness and depression take their toll. The weight of it all is just too much. But then, too, you have your pink moments of patience, understanding, and compassion. You have your golden moments of inspired joy and unconditional and abiding love.

In this state of inner conflict, the first center doesn't stop trying to protect itself just because higher consciousness has come into the mix of awareness. The consciousness of your cells continues to stay hard at work, doing what it must to keep the body safe. The second center still wants to belong and to be popular with the group in which it finds itself at any given time. You find that some of your social groups (maybe even your family) remain driven by ego, but new ones are increasingly governed by love.

The third center head is busy analyzing it all, wanting to achieve perfection, and is very judgmental of self and others in the process. The fourth center heart is aching for "everyone" (i.e., each of the centers) to simply get along and to figure out how to work together to get everyone's lowest needs and highest aspirations met.

The fifth center seeks authentic communication and self-expression for all. The sixth brings visions of the highest ways to achieve all tasks, if the third and lower centers would let it be. The seventh center, the point that serves to interconnect and align them all, whispers the answers, but they sure are hard to hear amidst the uproar.

The soul, collective consciousness that it is as center four, moves you to be love, to radiate love to all. Perhaps in ways you've never imagined before now. On the other end, centers one through three—all states of "less than"—the ego consciousness seeks to fulfill itself, physically, emotionally, and intellectually.

This is indeed the battle described by the Bhagavad Gita. This is the state of consciousness where both sides war for our attention. Angel on the right shoulder, devil on the left, as they say.

This is the way one fifty-year-old woman described her spiritual anxiety, her sense of duality, inner conflict, and whirling mixed emotions and motivations, having experienced awakening but still struggling to move into her authentic self as her dominant way of life. Personally, I'm grateful that she journaled her feelings and agreed to share them with you here, so that you won't feel so alone.

> I'm emotionally and physically drained. My anxiety level is at an all-time high. I'm feeling overwhelmed by so many emotions. To be honest, it's kind of unsettling, and I'm trying to keep it together. There is a convergence of thoughts, feelings, emotions, needs, wants, and desires that are all hitting me at once. I'm not sure what to trust. I'm breathing through the panic and crying through the sadness and regret, but it's tough. I'm drowning and I can't motivate myself. . . . I sort of feel like I'm losing my grip. . . . I'm lost and don't know up from down. I don't know what is real or what to trust. I can't even trust my own feelings.
>
> It feels as though my energetic systems are all haywire. And I'm confused as to what's going on with me. I've been keeping busy which helps . . . but there is this lingering cloud of anxiety. A kind of dissociative feeling and tightness in my throat and heaviness on my chest. It feels as if I'm going through the motions but not fully present. I'm depleted and drained.
>
> Just back from my family vacation, it was straining to be with everyone 24/7 and withhold my true self because

they just wouldn't "get me." I had an interior world that I kept to myself, which made me feel like a liar. Feelings and perspectives I couldn't share, so I felt detached. It was frustrating to be around such dysfunction and keep up pretenses. I feel this way with them almost all the time now. They don't get me or just don't want to hear it, so I keep huge parts of myself a secret. Which makes me feel like I'm floating outside of my life and not fully present and engaged. I'm not sure where I belong.

My life feels so out of synch. I can't even be my true self with the one person I'm supposedly closest to and spend the majority of my time with. This is why I'm not sure what is real and what is an illusion. Who am I if I can't be myself and I instead pretend to be someone I'm not? Who am I really? Who do I love? Who and what is important to me? It all feels so vague. I've held back, kept in check, my truth for so long that now I'm not sure what my truth is anymore.

Here is how a college freshman described it in his journal, before he had a name for the duality consciousness that then paralyzed him:

Why is it that I stay up late at night for fear that when I fall asleep, I will wake up to a day full of new opportunities? Why do I struggle to arise in the morning for fear that I will have to put forth most if not all of my energy and attention? Is this the definition of stress? A chronic state of unease, fearfulness of my own potential.

Why is it that those who have the capability for deeper thought must also possess the torture of self-conscious thought? Every action under judgment from my own harsh mind, determined to break the very spirit of the body in which it so comfortably resides. What can I do that will put both mind and body at ease, while also working to achieve mastery over both? To what ends must one go to reach a state of comfort and happiness?

I am tired of the cyclic nature of emotions. Too soon is the arrival of depression after happiness and vice versa. It is consistency in behavior that creates results, but behavior is controlled by state of mind, and thus my behavior is erratic and spontaneous. One day I will work with every ounce of my power, and the next I will take every shortcut and search for any way to accomplish nothing and still be at ease. What can I do to take complete control of my mindset, even during a most stressful time of my life?

Initial Actions/Tools for Coping

Note that a chronic state of unease is the result of chronic patterns of unconstructive thinking. Most of us have yet to discipline our thinking. Our mind is indeed like a monkey or like a new puppy, jumping all over the place and getting us into trouble, yanking our chain here and there. Our attention span, on average, tends to be less than eight seconds long. Our thoughts control us because our emotions go where our thoughts go, and our thoughts are usually all over the place. Leashed to our emotions, our behaviors are reactive, more instinctual than thoughtful. This is the neuroplastic normal, autopilot, habitual way of being. So below are some tools for disciplining your thinking.

- Center yourself. Focus your bodily awareness on your heart/abdomen area, whichever point feels most like the core center of your being. Breathe while focusing until you feel centered.

- Observe which lines of your thinking are constructive and uplifting and which are unconstructive, paralyzing, or even destructive. Even ten minutes of logging the unconstructive thoughts, followed by ten minutes of logging constructive thoughts as a daily evening exercise can do wonders, as the neuroplasticity of the body immediately begins adjusting to this new repeated action. For instance, catch beliefs that unloving or punitive self-talk is helpful. And stop it, the unloving self-talk, as soon as you catch your mind there. All it does is perpetuate a state of consciousness that is unloving, and keeps you even longer in the very state you wish to change. Instead, use compassionate self-talk. Talk to yourself as you would a good friend who is in need of your listening, your empathy, your understanding. Talk to yourself as you would a loving parent to a child. Gently remind yourself of your choices and the consequences they bring. In an observer mindset, be mindful of which choices are most soulful and healthy, which indulge the physical self without much harm done, and which choices are in fact unhealthy and destructive.

- Practice mindfulness meditation, where you focus intimately on the minute details of the focused object. Pick an object, even something as simple as a piece of fruit. Notice its texture, shape, scent, color, taste, and any other quality you can pick up with each of the physical senses. When you notice your mind wander, bring it back to the object. When it wanders again, bring it back again. Do this for two minutes to start, twice per day, and build up to five minutes each sitting over a couple of weeks. Notice the difference in your attention span the longer you practice.

- When you are overwhelmed in duality consciousness with conflicting sides of yourself and confused about which path to take when, make a list with two columns. In column A, list your various values, each in its own row. Note which center of consciousness it is stemming from. In column B, describe the consequences of taking that particular path, or not taking it, and whether the experiences they bring are constructive or not toward your spiritual growth. Allow these choices/consequences to sink into your awareness. Notice which path of experience pulls you the most. Decide from there, from a state of greater awareness and observer mind.

- To ground yourself with a renewed sense of reality, use the Consciousness Map to separate your ego's wants, needs, and desires from your soul's aspirations and highest inspirations. Then, as you would collaborate if you were at work in a committee meeting, attend to all of the centers. Instead of thinking "either/or," try to think of "and." Consider how you can keep yourself feeling some security for center one, *and* which things you can talk about or do together with your more ego-centered friends and family, *and* what you need to do to feel healthy, and what you can do to feel like you are thriving, *and* what you can do to feel like your soul is getting some authentic expression.

- What each of us needs to do when we're spiritually anxious and feeling this torn apart, more than anything else in my opinion, is to focus on "holding steady" our spiritual light. And whatever we need to do in order to accomplish that goal is what we need to do. Don't ruminate that it isn't your "whole" self 24/7. Practice, take baby steps. And focus on *which* aspects of your higher self are being incorporated into your life; that is, are being assimilated and integrated with the life you've been leading before your spiritual emergence. *And* breathe! Hold patience for yourself.

You literally have eternity to unfold your quantum nature.

3) You Make Ego the Enemy

Righteousness anxiety, I've called it. And righteousness anger, frustration, and so forth.

You have the thought that you don't want your ego to overtake your soul, so you take to battle to win a war you've waged against your very self. And oftentimes, not only do you bring this attitude into your life, but you also carry it with you as you judge all others.

Or maybe for you it's not that extreme. But you do notice that you judge the ego as not a good thing, both in yourself and others. Others may get angry at you because you keep pointing out "you're in your ego" and you proceed to list their faults.

But as we've learned the hard way, "a house divided cannot stand."

Initial Actions/Tools for Coping

- Take note that your ego, or lowest three centers of consciousness, is not your enemy. It is the only instrument for spirit that you have in this physical reality.

- Instead of making ego the enemy (and only ego would make anything an enemy), your new directive is to point your third center mental focus at center four and "hold steady the light" of love. Take responsibility for just one thing you can do to make the world a more loving place, *making use of the core qualities you carry as a soul* into this plane *through the use of your ego* or personality self. Align each of your centers so that your *lower are in regular service of the higher.*

- Practice catching your unconstructive self-talk, especially the soul or ego, holy or evil, religious or not religious, good or bad, friend or enemy, pendulum-swinging polarized forms of thinking. Then challenge yourself to think in constructive terms, ways that support, build, and uplift all *inclusively.*

- *Practice* allowing ego to serve the soul. Love who you can as often as you can. *Practice* inclusivity, patience, listening, kindness, respect, cooperation, collaboration, compassion, and forgiveness. For yourself and others.

4) You No Longer Fit In

Typically, you become frustrated that what you now believe, your new perception of reality, is beyond your ability to put it into words. And when you have tried to describe to others your experience, you have often been met with skepticism, teasing, ridicule, and even rejection, causing great anxiety about the future of your social and professional relationships.

Because of your new transpersonal, rather than personal, consciousness, you suddenly feel radically different from others in worldview, values, motivations, and many other ways. So different that you may feel you were beamed in from another planet or another part of the cosmos. (And what if that is true?!)

You worry that you don't fit in anymore, because from your past experience, the way others see the world is not how you see the world. Your new beliefs about reality can cause big problems between you and your friends, your partner, and your family, as well as your coworkers and the community at large. Still in their lower centers, they may snicker, call you whacko, weirdo, or accuse you of going off the deep end. They may pull away, no longer feeling they can relate to you. Or they may feel too scared to stay around for you, out of their own threatened sense of insecurity or vulnerability.

As an example, you've had an experience that all is love, that everyone and everything are one interconnected Wholeness characterized by a love that most fail to understand. You choose to live in accordance with that and are called naïve. Your business partner isn't sure she still wants to run a business with you. And your life partner insists you go to therapy to "fix" yourself.

Or you have come back from a near-death experience with psychic abilities, especially from centers six and seven, that you didn't have previously. Your spouse threatens to leave you, feeling too threatened and vulnerable in centers one and two to stay. "I have no more privacy around you, and I can't handle that" is the common complaint.

Initial Actions/Tools for Coping

- When you feel like you *no longer fit in* with the people in your life, it is especially the time for you to practice *discernment.* Instead of judging yourself and others in all-or-nothing ways (e.g., I do fit in or I don't fit in), discern. Discern ways you *can* fit in; that is, what you *can* talk about with each of your groups of people or what you *can* do to enjoy quality time with them. Your spouse, partner, friends, and family don't each have to have everything in common with you, no more now than before your awakening.

- Broaden the number of circles of people you hang out with, so that you have a circle for each aspect of your being, both egoic self and soul self, if that is where you are in your journey. Use social media and Internet resources like Meetup.com to find new groups.

- Search under keywords that you find very important to you and notice what types of meetings, conferences, and events come up. Pay attention to where they are held, for there are certain locations in the US, like the Boulder-Denver area of Colorado, that are known hubs for attracting people of higher consciousness. By doing this kind of search, you will learn about many inspiring people who are already out there with similar thinking as yourself, and it may just give you courage to join them!

5) Career Confusion

Your awakening has left you confused about what to do for your career. You feel inspired toward some new endeavors but fear making the changes that seem so big right now. You especially fear that you can't make a decent living doing your soul work, and that terrifies you.

Or, in all-or-nothing thinking, you tell yourself that money is evil, yet fear living without it.

You may tell yourself that you just have to keep plugging away at the job you hate, because it is the responsible thing to do as a spouse or a parent. But your mood is eroding and beginning to affect your relationships and your health. You're anxious and frustrated, feeling stuck in the career path you chose, maybe even still facing student loans.

Or you don't yet know what your calling is. You've tried all different kinds of jobs, and it seems like none of them lasts for very long. None of them has really brought you a sense of meaning and purpose. And it appears as though there is always some huge block in your way. Financial expectations that don't pan out. Bosses who seem to have it out for you. Coworkers who just don't understand you. Yet you know that you work really hard to try to make the job succeed.

Many of you are young and just starting to think about a career. What you long to do seems like it doesn't exist, but you feel so deliciously alive whenever you think about it. And then you face the job search process, and your energy falls flat because nothing you see feels exciting.

Oftentimes, we are trying on for ourselves what seems to work for others. But we are not being present with who we are inside, or we are ignoring that aspect of self out of fear. A young yoga instructor was fretting about what to do next in her career. She felt a dreadful sense of urgency to hurry up and pick something at which she could "actually make a living." She didn't believe she could earn a living from being a yoga teacher, so she was pushing herself to figure out a plan B.

She had been making only about $30,000 per year, but it wasn't an issue for her until she got divorced and had to rely on only her income. Her parents were glad to help her get settled into her new life, but she felt like a failure for needing their help.

As the days passed by, she began having panic attacks, unable to decide what else to do for a career. The fearful ego was clearly blocking her sense of soul purpose.

That's when she came to see me.

One of the things that together we helped her see was that she was punishing herself with her self-talk, believing that would motivate her to move faster. But all it did was lead her to dissociate or numb out with alcohol in order to avoid panic.

We helped her to instead practice constructive self-talk.

And we had her center herself in her joy, which was yoga. From there, as part of an experiential writing and imagery exercise, we had her write down every vision that arose within her mind as she focused on being immersed in yoga. She wrote in the present tense, seeing it has happening in the now. Statement after statement, she wrote *as she received.* And as she did, her mood shifted into excitement.

As she maintained this imagery through her sixth center consciousness, the visions shifted from seeing herself teaching yoga at someone else's studio to opening her own studio. Her own studio transformed in her mind's eye into a healing center, then into a community center that was a robust sanctuary for conscious and sustainable experience, healing, and livingness. And it held multiple opportunities for a variety of income streams so that she didn't limit herself to just teacher pay.

Fear arose in her lower centers from seeing how big her vision had grown. So we had her write down various baby steps to do as homework, such as research local healing centers, eco resources, funding sources, and like-minded others who were eager to join in such a vision. We had her focus on one step at a time and on remembering that we don't create these visions alone. We do it together, as a "we," with similar minded others.

Career confusion is a sign that we have yet to align our vocational life with our higher centers of consciousness. Soul energies are trying to come in, and we are tuning them out.

The longer we neglect those energies or streams of consciousness trying to come in, the more physically we feel the mounting pressure in our body. Eventually, the subtle energy pains will transmute to physical disease if we don't pay attention and make the life changes we are moved to make.

Initial Actions/Tools for Coping

The key in resolving career confusion is to first know yourself, in all centers of consciousness to which you have access and awareness. Allow the higher centers to inspire your heart, the barometer and compass for your soul in the body. Notice which ideas for career are energizing and heart expanding. Allow the head to do its part by suggesting ways to get to where the heart is leading you. Let your mental and physical activity in the personality centers serve as the vehicle of your soul, and you will inevitably find the career happiness you've been hoping for.

- For help on this, do this PEMS exercise. Take out a piece of paper. Divide it into four sections: physical, emotional/social, mental/intellectual, and spiritual sense of self (PEMS). Make note of what enlivens you in each of these four aspects. List qualities, traits, skills, strengths, interests, what energizes you, what brings you peace and joy. Notice how there may be a pattern to what naturally lifts you into your most spiritual high, the ideas or topics that most inspire you and even make you feel giddy inside, and the activities that make you feel healthiest when you are engaged in them. Out of this pattern are likely your clues as to your soul's preferred career direction for you, if you have yet to discern it through your intuition.

For a very short example, physically you may thrive by actively directing a nonprofit on a daily basis. You have the physical health and stamina to do so, and the physical resources. Your teaching degree is in hand, as is the experience you need to do what you dream. Emotionally/socially, you may feel passionate about teaching children. You have the skills of patience, compassion, and collaborating with others. Mentally, you may love to communicate with others about better ways to educate children so that they are loving and consciously creating in their own lives. You are naturally a superb teacher. You love keeping up with all things educational. Spiritually, you know you have been called to uplift others, children especially. You feel alive each time you imagine opening the nonprofit.

- Do a "tree exercise" for yourself. We can think of a tree as a metaphor for the vocational or career self. The roots of the tree can depict the key qualities within us from Source that we became aware of in our past, even in childhood. The trunk of the tree represents the present state of self, those qualities within us that now feel important to us to create from. And the branches of the tree hold an array of future probabilities, each vocational aspiration given a place in our potential or probable future, so that we may look at our various inspirations from an eagle's eye view and see which may combine into a uniquely creative career that is needed from you by the Universe.

- Holland's RIASEC (Realistic, Investigative, Artistic, Social, Enterprising, Conventional) exercise, though traditional, is still a helpful career tool in my opinion. I wish all high school freshman were required to find out their top three personality codes of the six. It is useful to understand which are our top personality preferences when it comes to careers, and which careers exemplify each of the six personality codes.

The main limitation to the testing process, perhaps, is that it doesn't take into consideration the soul and how the soul may move us to create that which doesn't currently exist. So when I took the test in graduate school, the results suggested that being a "fine artist" was my optimal career; this despite the fact that I had never cared to dabble in paints or art projects, unless I had to for school, and had no current interest in doing so. However, coming away with the insight that my top two personality codes were *AI*, with *ES* fairly tied after those, was very helpful.

Holland suggests we test for our top three codes and compare them to those required by the careers we are thinking about. If you are strongly "artistic," then you will clearly be miserable in a job that punishes creativity and demands that you do the tasks in a very specific way. If you are very "social" and you attempt to do a job that requires sitting in front of a desk all day, talking to virtually no one, then you are not putting yourself in a job that aligns with your skill set. And you will feel like a failure when you regularly don't fit in.

6) Spiritual Ambition

You've experienced an awakening into your higher spheres of consciousness. And you are moved to align your life with centers four and higher. You know what your soul begs for you to do, and you're taking steps toward these higher aspirations.

But along the way, it's hard for you to distinguish the voice of your soul from the voice of your ego, because the latter has been the basis of your identity for your entire life until only recently.

Remember, the soul feels aspirations; the ego feels ambitions.

I call it spiritual ambition when the ego is more in charge, and our actions are motivated really more by financial profit or how we can win or protect ourselves personally than by what's in the best interest of the mission. It can cause problems, because it can sidetrack us from being in synch with that Great Conductor, can cause us to derail altogether or, worse, cause a "contamination spill" as we leak our ego stuff and darken the purity of the original spiritual mission. Here, the ego deludes us into thinking we are "more spiritual" than we really are. We make career moves that are not part of our soul path, then wonder why "God abandoned" us by not helping us more synchronously.

The way it felt for me initially was that I was very clear that the soul intended for me to work in the field of psychology; however, ego had its own ideas about how that would be done. My ego, all three lower centers, was determined to complete a doctorate and practice "more safely" as a psychologist. My analytical mind, center three, believed that the conventional PhD was the way to go to minimize the ridicule around the spiritual direction I was now headed.

Center two, the part of me that worried about fitting in, had mixed feelings, preferring to go to a transpersonal program that would "get me" but influenced by center three that an APA-accredited program would be more strategic. Center one wanted whatever felt safer, more in control, and above criticism. I was literally torn within about which way to go about becoming a holistic counselor.

The way it worked itself out was not so even-keeled. Despite being synchronistically told to "get out of school sooner rather than later," I still pushed for the APA-accredited PhD programs but did not get accepted, no matter what I did. While interviewing for the PhD program at USM, they asked me if I would consider switching my application to the master's program, so I did. While in the program, which felt painfully medical model in flavor, I was blessedly led toward work in another department with a Native American professor who wanted to pursue research along more spiritual identity and moral development lines. Thank God! Literally.

When I went into Dark Night of the Soul, it was she who guided me out by sharing her St. John of the Cross book with me. And it was she who shared a soul dream with me, which I recognized as a message again to get out sooner rather than later. So, instead of accepting the offer to transfer into the PhD program at the end of my master's, I declined, moved as intuited to Colorado, started an Independent Living Program for a residential treatment center while working toward licensure, and finally moved into private practice when Spirit urged me forward.

Clearly, Spirit knew there were many of you out there hoping for someone like me to help you. Spirit knew a PhD likely would take me further away from the direction in which it was leading me. But I had no idea, for this kind of work was unknown territory. This is why we have to trust that Spirit knows best when it sends us intuitive and synchronous messages, and why we "Let go, and let God."

Initial Actions/Tools for Coping

So when you are in the process of changing your life so that it aligns with your highest self, be on alert that your motivations remain pure.

- Check in regularly. Notice which are ego's pushes as compared to soul's pulls. Remember, ego seeks to achieve as an individual and to be seen as better than others. On the other hand, feeling responsible to do its share, the soul aspires to lift up and collaborate with all others—inclusively, patiently, and compassionately—to make the world a more loving place.

- Ask yourself, "Is this to build me up personally? Or to build up others?"

- And practice, practice, practice "holding steady the light" of pure soul. Let it become your life's minute-by-minute meditation. When you catch yourself in ego, don't judge, simply redirect toward the voice of your soul.

7) Pure Soul

For the more pure soul, who has been without much experience in the physical plane and *in touch with Wholeness from birth*, being in this physical dimension can be very painful.

You don't like the material sensations of center one, the sense of distinct boundary, or the feeling of separateness from others in center two.

Here, in the physical dimension, you feel so isolated and alone as compared to the interconnectedness you remember (from your holographic mind). Others' energies feel so dark, fearful, and angry. It hurts to feel them. Their energy overwhelms you at center two.

The physical body feels painfully dense and heavy. The impurities in food and other pollution make you sick.

Compared to the higher centers of consciousness, you find that your lower centers are slow and muddied with illusion. It feels like being trapped in a circus hall of mirrors from which you have no immediate escape.

Your fourth center heart beats for loving collaboration with others. But everywhere you turn, you're surrounded by a sea of others who are centered much lower in the vibrations of hatred, violence, and competition.

Your mind doesn't understand them. You may even find yourself avoiding the pain of being near them and withdrawing into your own world of refuge. But that strategy does not feel in alignment with your higher nature, and the anxiety worsens.

You struggle to find and balance a job that you can live with, one with values that align with yours, matching you in every center. You keep searching and not finding work relationships that resonate with you. The subject matter that interests you seems to be above the consciousness of the people you are around. You get that "deer in headlights" look anytime you try to converse about the topics that really move you in center three, the service that moves you in four, the career or creative expression that ignites you in five and six, and all of that which feels completely authentic and purposefully inspired in center seven.

You grieve for the Wholeness, *long* for it. Sometimes you cry about how much you miss it. Secretly, you pray to be taken back to Source.

Yet you're here to do the job you came for, which is to anchor a much higher consciousness and vibration into this realm. And you're determined to hold true to the heartfelt responsibility for which you, at some higher vibrational level, volunteered.

One of my clearly pure soul clients seemed most helped by the Consciousness Map, because it gave her a way to organize her thinking about the types of people she encountered. She asked for some basic skills and "street smarts" (as she phrased it), because it was as if she were a clean slate and had no experiences from which to draw upon. She asked for help with "standing up for herself," for dealing with anyone who wasn't naturally fair, loving, and collaborative (which terrified her).

I taught her how to discern her skills, how to engage assertively with others, how to do a résumé, how to seek and accept work that aligned with her loving nature instead of thinking she had to change herself to fit in. Overall, she needed the skills to fit in with the basics of this society but the mastery of not losing herself in the process.

Initial Actions/Tools for Coping

There are some basic things you can do if you are one of the pure souls who is on the planet now to help humanity evolve.

- First, attend to your physical health, center one. Make sure to get adequate water, fresh fruits and vegetables. Your light body likes to eat light, in both senses of the word! Some of you will feel better on a more liquid and plant-based diet; however, others of you do better with meat in your diet because it helps ground you. It lowers your vibration, but in a constructive way for you.

- Get plenty of movement in your day so that you can purge yourself of dark energy or energy that is not resonating for you.

- You can also do imagery exercises to circulate dark energy out and bring pure energy in. Breathe out the dark polluting or stagnant energy through the soles of your feet, then breathe in the light through the crown of your head. You can use white light or intuit which color frequency you need in the moment.

- Dare to reach out and find others who share your values, for there are many more pure souls now than you individually realize. The main goal is to balance all the centers so that they are in alignment with one another as best they can be in this physical dimension.

- Allow yourself the conditions in which you thrive at every center of your being.

8) Extreme Dissociation or Feeling Ungrounded

Retreating into your higher frequency centers too often day to day can cause you to come across as spacey or ungrounded to others, especially if you've become addicted to your meditative states.

You're enjoying the higher states of consciousness that you've learned to access through focused intent and don't feel like coming off the natural high. You have a tendency to ignore your more mundane responsibilities, and it causes problems in your life. Especially if it causes you to stop functioning, to experience difficulties at work, to lose your significant relationships, and worse.

Or you experienced trauma as a child and coped by going out-of-body at will. Now an adult, you have over-generalized this coping tool, and it has become an unconstructive habit.

You are clearly and significantly out of balance, favoring the higher and neglecting the lower physical aspects of self.

Initial Actions/Tools for Coping

Balance is what you need to prioritize.

- First of all, *ground yourself* in center one. See your physician (preferably vibrational medicine minded) to evaluate your needs for medication (if you are clinically and significantly not functioning or are experiencing psychosis) and nutrition. You may be more sensitive than the norm to any substances, so medication, and even supplementation, amounts are important to monitor during this phase of your shift.

> Penny Montgomery, PhD, and her trained staff in Denver, Colorado, can detect with neurofeedback those who have brain wave patterns that are more sensitive to substances, including prescription medicine.

- Make sure you *stay off recreational drugs, even marijuana,* which is not constructive for energy sensitives particularly at this time, even if it is legal. My clients who are sensitives tend to experience paranoia and end up hospitalized for acute psychosis when they use marijuana, especially when they use it chronically, and even if they are using it medically. It only temporarily solves their anxiety problems, and it mostly causes even more issues to deal with.

- Instead of self-medicating for your overwhelming energetic feelings, consider working with an energy savvy counselor or psychotherapist who can help you learn healthier ways to cope with transpersonal anxiety and other uncomfortable emotions that may arise. There are also those who can teach you how to transcend at will in a healthy manner, one that uses your focused consciousness and does not need substances; but wait until you feel more grounded.

- Be proactive in your coping strategy by adding lots of soulful activities to your physical day to day that keep you stable and functioning in center one; it helps you maintain a more steady and predictable mood in center two. Then when you do need to add intervention tools to cope with overall stress in the moment, it won't feel nearly as difficult.

- Practice scaling your emotions from 0 to 10, with 10 being the most intense mood. For instance, if you are feeling anxious, notice when you are in the 2–5 range; that is the time to use a coping intervention tool. It is way harder for your coping tools to work when you shoot above 5. Practice regularly mentally scaling your emotions in this way.

- If you are feeling ungrounded, *spiritual yoga or similar practices may not be the most constructive for you at this time.* Not if they are intended to help you expand into transpersonal awareness. What you *do need, is somatic or body mindfulness,* where you attend to the center one *physicality* of your body, such as your body in your chair or your feet on the floor. Feel your feet connecting to the floor or earth. Feel your skin and your boundaries. Now is not the time to dissolve!

9) Wondering "Am I Crazy?"

Perhaps the worst of all spiritual anxiety is when you wonder if you are crazy. Or maybe it's even worse when you know you aren't, yet face others who want you diagnosed with some psychotic disorder and hospitalized "for your own safety."

In my experience, if my clients are wondering if they are crazy, they usually aren't.

In psychology, we don't really use the word "crazy." We do diagnose psychosis and conditions like schizophrenia. The two distinguishing features of psychosis are hallucinations and delusions. If a psychotherapist believes you are sensing something that is "not real," he or she will say you are hallucinating. If he or she believes you believe something that is "not real," you will be noted as delusional.

The way I differentiate psychosis from "quantum" or higher consciousness is fairly straightforward. If you can maintain a conversation; that is, if you are able to track and understand what I am saying and respond in a way that seems to flow, then I don't diagnose you as experiencing psychosis just because of the content of your belief or sensory system.

If you can't track and communicate with me in an internally coherent, logically consistent way; that is, if you wander off too far, or if you seem completely unaware and unresponsive or in a world of your own, and you can't engage with me in a two-way conversation, then I know that there is something else going on. Psychosis is just one of those things.

An example in which I referred someone because I did in fact suspect schizophrenia was when a woman brought in a neighbor boy. He was about fifteen and seemed vaguely aware of what I might ask of him. But it was like he would hear one cue word from my question and wander off on some monologue that was hardly in line with the question. There never was any conversation, because he couldn't track in that manner.

In my opinion, symptoms like this, suggestive of schizophrenia, can be a classic and worst-case effect of ungrounded transpersonal awareness, especially caused by damage of the physical brain, the antenna needed to transmute and translate the higher vibrational energy into the lower. The damage to the antenna brain may come from many causes; but many of us transpersonal therapists note how often the use of drugs seems to prematurely "tear the etheric/energetic/blueprint layer" that is naturally in place to protect the human psyche until it is naturally ready for higher vibrational spheres of consciousness.

A few times, I have had clients come in that were pretty high functioning for schizophrenia. Before immediately referring them, I gave them the benefit of the doubt when it came to helping them work toward the "I want a job" goal that they listed. But in each of those cases, I was readily able to see that something else was going on clinically as soon as I saw that they couldn't write in any linear or logically consistent and coherent manner. They couldn't even begin to write a résumé of their work history, the very first homework I assigned. Words and sentences were extremely disjointed, barely recognizable.

A man in his forties had been using a long list of drugs, as well as alcohol, since he had been in middle school. He had just been released from successive stays at two area hospitals, where he had been diagnosed as having had a psychotic episode due to marijuana use. The reason he was in my office was because he'd promised his wife and friends that he would come.

According to how he felt personally, he didn't believe he had a problem. The way he described it was that the combination of drugs he had recently taken at an overly packed awakening concert had actually led to his awakening. The language he used was all about love. He had felt Love. He knew he was here to be Love, to radiate Love, to teach Love.

How could I argue with that?

The problem, it seemed, was that the way he spoke after his awakening was so different from the way he had spoken previously. It was causing his wife and friends to withdraw from him a bit. Clearly they thought he was speaking like a "crazy" person, especially when he talked about his encounters with shapeshifters and beings from different planets. (I told him to check out Dolores Cannon's books and to know that he is not alone in his experiences.)

He was apparently now on a different wavelength. Literally. There was a disconnect between him, in his current state of consciousness, and his friends and family in theirs. I was able to use the Consciousness Map to clarify that for him and to teach him how to be judicious in his conversations (e.g., how to discern to whom to say what and when and which language to use depending on which centers are predominant for a particular person).

Without quantum physics to explain the possibility of parallel realities (i.e., universes or multiverses, as they are starting to say), or multidimensional frequencies, or other dimensions of reality existing at different bands of frequency or speeds of vibration overlaying our spacetime and even beyond spacetime . . . what chance did he have at *not* being called "crazy"?

My client didn't seem to be in a psychotic state in my office. But I could tell that I had to help him understand the experiences during his high that were in this physical dimension and how to differentiate them from those that may have been in other dimensions of frequency. For instance, in this dimension, he was taken to a hospital via ambulance. But as his consciousness traveled beyond, it was clearly in a different frequency of existence that he received guidance from the shapeshifting shaman he saw sitting in his hospital room chair.

Perhaps we could say that his consciousness may have "accidently" collided with some other or parallel dimension when he saw people in his hospital room that were supposedly not really there. (Such a collision may also explain why "crowded rooms" are one of three most common deathbed experiences; why planes suddenly and mysteriously disappear, especially around the Bermuda Triangle; how something like the Philadelphia Experiment might actually occur; how Dr. Harry Oldfield can capture people, furniture, and mausoleums from other times with his specialized equipment; or how scientists are now finding exoplanets invisible to our eyes.)

Maybe he was simply tapping into another frequency reality overlaying this one. What may seem like a hallucination to us may simply be experiences of some other dimensions of frequency or wavelength of energy, some "spooky action" in a nonlocal parallel universe.

People who have awakened transpersonally are no more delusional than anybody else. When it comes right down to it, we all have beliefs with no basis in fact, including doctors, psychiatrists, and psychologists.

When it comes to delusions and hallucinations, given our new knowledge of quantum reality, we therapists really need to apply the above criteria before misdiagnosing someone as psychotic who is really just psychic and who really can manifest with their energy, even if we can't manifest with ours.

Initial Actions/Tools for Coping

- Again, if you are wondering whether or not you are crazy, the fact that you are wondering is likely a sign that you are not. On the other hand, see a transpersonal psychotherapist to help you through this distressing time. He or she does understand the difference between mental health disorders and spiritual or transpersonal emergence and the anxiety it causes. Such therapists have been trained in the matter and have usually had their own experiences with the issue of transpersonal consciousness.

- Later, when we talk about being psychically overwhelmed, we will more fully address the goings on when someone encounters ghosts or foreign entities. For that, too, is definitely one of the big sources of transpersonal, I'll even say quantum, anxiety. Just as for my teen, who ultimately became clinically depressed as a result of the despair from such long-term anxiety. In his case, he was scared that he was going crazy after he began to see ghosts and afraid to let anyone know about his psychic abilities for fear of how they would respond to him.

- Ground yourself, using many of the tools already discussed regarding foods/nutrition, water, sleep, planting your feet in the soil, and particularly body mindfulness.

Chapter Four

Coping with the Dark Night of the Soul and Transpersonal/Spiritual Depression

As with the previous descriptions of spiritual anxiety, spiritual depression feels like, and can be like, regular human depression, but its root cause is always related to a perceptual disconnect from Source and Love energy.

The Spiritual Dark Night of the Soul

You have long felt the presence of Spirit, but suddenly it has vanished. The depression you now feel is what St. John of the Cross called the spiritual "Dark Night of the Soul."

In my case, I felt as if I had been abandoned naked in some distant and foreign cold, dark desert at night. I felt alone, shivering, and with no bearing as to where I was or for how long I might be left there. I had never been without Presence before. This was new and unexpected. At the time, I had no idea what I had done "wrong" to get me to this place. On the contrary, I felt as if I had finally been doing "right" by following what God had asked of me to do.

For how long would I be deserted? I sobbed with despair. My heart seared with anxiety, disappointment, grief, frustration, and anger. I felt lost and confused, overcome by a sense of helplessness I had never known. I yelled at God, wanting to know why I had been abandoned without warning, without trace or clue as to how to get back to the feeling I had trusted more than my own heartbeat. Ignored, no answers coming, I stood face to face with the thickened fog of my depression.

With nowhere else to turn, I tearfully admitted what was going on to my research professor, a Native American woman whom I had come to trust with my secret spiritual life. She and I had shared many a dinner conversation over our similar spiritual quest, and it was she who suggested that I read "Dark Night of the Soul," a group of stanzas written by St. John of the Cross, the sixteenth-century mystic and contemporary of St. Teresa of Avila, who also wrote about the mystical journey of the soul in her work *The Interior Castle*. For those of you who seek a deep understanding of the Dark Night, I recommend that you read St. John's poem and the helpful translations and commentaries that are available to you, even for free online.

What I learned, in short, is that the spiritual Dark Night is a *milestone on the spiritual ladder,* one that comes when we have *already shown that we can remain pure-hearted in the midst of the chaos* of this world. It is a *time for us to test ourselves and see whether or not we can "hold steady our light" even without the tangible help from Spirit.* It is much like we have been riding our bike with the training wheels on, and Father has taken them off.

Of course, it can be a terrifying experience, and it seems to last as long as necessary. But from what I have come to know, the spiritual Dark Night appears to come to an end as soon as we let go of trying to get Presence back and as we steadfastly proceed forward with our spiritual mission, seemingly, and therefore bravely, on our own. We must learn to "walk in the dark," and as soon as we do, Spirit returns full force, but in another capacity that feels more cocreative or cooperative than before.

Recently, a client described feeling "so abandoned" by everyone. So tired of being there for everyone else and not feeling it reciprocated. So tired of the "lessons" that she wanted to punch the next person who pointed them out to her. So overwhelmed by the blows she was receiving from life—including having been hospitalized twice for debilitating flulike symptoms in the previous few weeks, with none of her friends reaching out to help—that she couldn't take it anymore.

It felt to her that she was completely alone, without any control or effect in the world, because no matter how hard she tried to find a sense of community with loving others, it didn't happen. No matter how many less-than-perfect people around her had plenty of intimate relationships over the years, she was always alone. And when she would meet someone and get her hopes up because of some amazing stimulating conversation, they would seem to vanish and not be heard of again.

Not that she didn't appreciate her customers and colleagues; her work was going very well, and the money was the best it had ever been in a single month and just in time for when she was so sick. But what she wanted most of all for her personal life, a deep and long-lasting boyfriend, partner, or spouse, she could not make happen. Loving reciprocal relationships just seemed to elude her. "Whatever . . ." she muttered.

In fact, as far as she was concerned, she didn't even believe in God or the Universe anymore. In this state of hopeless depression, she could no longer see evidence whatsoever of a conscious, loving, interactive, synchronized "Universe" at work in her life. And she was angry that she couldn't even say that to the people in her life without judgment, without them responding "I'll pray for you" in what felt to her like a pathetic, patronizing tone.

Clearly, in her thinking, God did not exist, because her prayers were not being answered. As I listened, I told her it sounded like the Dark Night of the Soul, the feeling of being abandoned by God or the Universe, the distressing state when you can't feel Presence anymore, the time when you come to doubt that God or such even exists, and your complete worldview seems shattered. There you are feeling like a limp asparagus, drained of life. And you don't care. Nothing matters. Might as well give up or maybe just go through the motions. Not necessarily suicidal—that would take energy—but "whatever . . ."

I asked her how she would draw what this Dark Night felt like to her. She responded, "I am just a tiny black dot surrounded by nothing. But the nothing feels white, because black feels like something." As she said this, it dawned on her that the white was indeed a blank canvas on which she could create a something of the nothing. It led her to say aloud all of the ways she planned to give her life meaning, by traveling and experiencing new cultures, by focusing on her clients and her work, and by living life on her terms until the very end. And it led her to exactly where she needed to be for the duration of the Dark Night; that is, *present, allowing, and open to letting in whatever* appeared.

Initial Actions/Tools for Coping

- Knowing that the Dark Night is a significant milestone in the journey of every soul, by itself, can feel helpful. It means you have been able to hold your light steady, at least to some degree, and are ready to learn to cocreate in a higher manner.

- If you are struggling with the spiritual Dark Night, be mindful of how your ego may have been trying to seize control of your spiritual path. Practice letting go, staying in the present moment, and alertly waiting for the new set of forces (that is, the highest centers) to move you, even though you probably can't feel them right now.

- Think of the movie *Star Wars*. Obi-Wan Kenobi taught young Luke Skywalker how to awaken consciously to the Force. Luke's heart was generally in the right place, but his ego often took over. A deeply symbolic scene, it was when repeatedly practicing wielding his lightsaber blindfolded that Luke learned to become one with the Force and was able to successfully navigate his spiritual disconnect. The

Dark Night is your time of training, of going it alone and "blind."

- The spiritual Dark Night is for a purpose, even if only the Divine knows what it is. In the example of Luke, his time of training blindfolded prepared him to go up against all odds, all alone, and achieve the impossible in the trench run in the Battle of Yavin. As the spirit of Obi-Wan reminded him, Luke used the Force instead of the targeting computer to perfectly shoot his torpedoes down the two-meter-wide exhaust port and successfully destroy the Death Star (another great symbol for the veil of illusion). Only because of his training blind was he able to eventually fulfill his purpose as a Jedi.

> Size matters not. Look at me. Judge me by my size, do you? Hmm? Hmm. And well you should not. For my ally is the Force, and a powerful ally it is. Life creates it, makes it grow. Its energy surrounds us and binds us. Luminous beings are we, not this crude matter. You must feel the Force around you; here, between you, me, the tree, the rock, everywhere, yes. Even between the land and the ship. —Yoda, *Star Wars: The Empire Strikes Back*

- To use another metaphor for describing the spiritual Dark Night, and how to navigate through it, it is like you as ego have been used to being the one driving your car (aka your body, of course). Then you begin to feel the Force driving the car through you. Eventually you are tested, in the spiritual Dark Night, by having to drive the car blindfolded with only the Force within to direct you along the way. It is in this phase of your spiritual training that you learn to become one with the Force, to trust its guidance so completely that you let the Force alone guide your way. As you do, the Dark Night seems to lift.

- I recommend that you try to draw what the Dark Night feels like to you. Often insight comes as you meditate on that image. By being open and allowing of "whatever" shows up, it seems to train us to live more intuitively and in concert with the synchronized unseen. And as we do, Presence returns.

The Dark Night of the Senses

Much earlier on the spiritual ladder is the dark night of the senses, lowercase letters, as St. John of the Cross differentiates it from the spiritual Dark Night, capitalized. It is certainly a time that feels like clinical depression, when nothing quite satisfies the senses. Pleasure from material sources seems to have disappeared.

In my modern-day interpretation, I would say that this dark night is a stage in which our consciousness is beginning to withdraw from the lower, more physical centers but has yet to anchor in the higher. It truly does feel like a place of purgatory, a painful state stuck between earthly and heavenly realms.

This is a state that often precedes your quantum leap, where, like an electron, you are pulling out of one orbit but have yet to emerge in the new.

In this place, you are ceasing to resonate with what used to give you pleasure, satisfaction, or contentment. But you have no idea what to do to feel happy again. Everything you try fails to sustain your attention or excitement for long. You feel a bit hopeless about ever feeling good again.

It is as if the soul knows that your physical senses are getting in the way of your evolutionary progress. And it decides to take away the obstacle, to give you a chance to move further forward. A different kind of blindfolding, but in this case all the physical electromagnetic senses are affected.

At the level of the cells and their neuroplastic normal, stimulation through material means has gotten old. It just doesn't do it for you anymore. It's like "been there, done that." But you are so used to getting pleasure from physical, especially chemical (natural and manmade), sources that you have no clue as to what else to turn to for a high.

The *point to the dark night of the senses is to move you.* To move you from the gestalt view of the dark vases and adjust your eyes to the view of the light faces (just as in the famous drawing by psychologist Edgar Rubin, where, depending on your focus, you can either see two profiles of people, or a vase between the two). And the ways to do so are as individual as you are. But they inevitably involve learning to perceive subtle energy and the intuitive information it brings.

Initial Actions/Tools for Coping

- When you are in this kind of depression, it is time to turn to that which is more subtle energy—Consciousness, Presence, the voice of God, the flow of the Universe, Source—however you choose to label that which is in reality beyond any sort of label.

  To start, you can *turn toward* in prayer and appreciation, learn to meditate, contemplate, or even take a class in perceiving subtle energy, such as an energy healing class, whether clinical qigong, Quantum Touch, Reiki, Therapeutic Touch, Healing Touch, Barbara Brennan or Donna Eden methods, or whichever moves you.

- You can meditate on feeling *gratitude* for even the smallest things you usually have taken for granted, like a rosy sunrise, a double rainbow emerging after a storm, a rejuvenating walk, the unfolding leaf of your houseplant, the dew on a blade of grass, the sound of wind chimes or the birds chirping. This is a kind of mindfulness meditation, where you pay close and intimate attention to the details of a single moment, and it anchors you into the present, where you have to learn to be if you want to get out of this depressed state.

- Notice that when you feel depressed, you are likely focused on the past; when you feel anxious, you are probably focused on the future. But when you are attending to the present moment, you are grounding yourself in the stability of the now. And the now is where you need to be in order to feel the Presence that you seek, as if it weren't already within you.

- When you are present to the subtleties of the moment, it is a good time to reach out and ask the Universe or God to come into your awareness. It is a ripe opportunity to ask open-ended questions and to be completely open to receiving an answer. And to allow the answer to come in through a variety of ways. Maybe in the passing comment of a stranger, a book that jumps off the shelf at you, a bumper sticker on the car in front of you, or a song playing on the radio. Why? Because you haven't been used to dialoguing directly, and the Universe knows when you need a mediator.

  Remember, it's *because we do have free will that we have to ask for what it is we want help with,* like answers to questions such as "What is my highest purpose? What am I here for?"

- If you want to practice conversing directly with the Universe, God, or Self, try this exercise I show to clients in my office. I refer to it as a *stepping contemplation.* Stand up and be prepared in a moment to take several steps from where you are currently standing. Close your eyes. Take a few relaxing breaths and bring yourself into the present moment. Feel yourself as soul centered, aware, open, and listening for what could come.

Now ask the Universe an open-ended question. If it were me doing this, I would probably simply ask, "Now what do you want me to do next?" And I would sense my answer. For some, the answer comes as words that pop unexpectedly into the mind. Others see pictures or images that contain the answer or direct the way forward. For you, notice what pops into your awareness and consider how that might be an answer or part of an answer.

When you feel like maybe you've received some sort of answer or clue, take a step. That step signifies to the Universe, "Okay, consider that done!" Then ask a follow-up question, if needed, or a "What next?" type of question. Continue to ask, receive a response, take a step as a sign of action, then ask another question. Repeat.

Some of my clients may be too self-conscious to do this at first, especially in front of me. But I can usually tell when they are ready. Sometimes they sit down sooner rather than later, because the rapidity of the response tends to freak them out a little bit. But others enter a totally new realm of reality, because they quickly discover a Universe that really is waiting for their coparticipation. It is *as we act, that we receive* what we need from the Universe, so in this exercise, the stepping serves as the action. And it is interesting what we receive when we are in action, in the process of moving in concert with Spirit.

Just recently, one woman shared a few of the several images that came to her during this exercise . . . such as a bird flying, a telephone, and a crown. Because she naturally intuited the qualities she associated with each of these symbols, she came to understand the Universe telling her to let go and fly, to do her soul work, then to share it with others, and eventually her path would lead her to feel the ultimate power of Self victorious.

I am often asked how to know whether the answer came from the brain or from higher sources. My answer is this: if it is you or your brain chatter, it will feel like you "did" something; if it is intuited from higher realms of consciousness, it will feel like you "received" something. Just like when you are talking to a friend; you easily know the difference between what you say and what you hear.

Feeling Like You Are Not Living Up to Your Spiritual Aspirations

You judge yourself for not living up to your spiritual aspirations. You feel like a disappointment or like a failure for not completely aligning with the perfect picture you hold of what your life "should" look like as a spiritual person. You feel guilty or ashamed for your "imperfect" ways of being. The depression is mounting.

Initial Actions/Tools for Coping

When you judge yourself overly harshly, you will make yourself depressed. Why? Because your ego mind truly is de-pressing (in that it is depressing but there is also a downward pressing motion) and suppressing the literal energies of your soul voice, until you can no longer even hear it.

- When you are in this state, it is time to become aware of your thoughts. Time to notice which are constructive, encouraging, and supportive. And which are not.

  For instance, toward the end of your day, take ten minutes to list the most common thoughts that arose throughout the day. For every thought you realize is unconstructive (e.g., you're a failure, you're bad, God would be disappointed in you), write down a constructive one (e.g., you have completed this task, you have become good at doing this for others, you are learning to be more patient, you did practice kindness today). Make these thoughts about the baby steps you *can* and *have* taken toward your spiritual aspirations, mission, or purpose.

- If you need to give yourself extra-credit homework, then also write down next to each thought the emotion you feel. Soon you will learn which thoughts spiral you into your depressed mood and which thoughts uplift you. Practice the latter! Whenever you catch yourself in unconstructive thinking, switch to those thoughts you know uplift you.

- Remind yourself that all of us are works in progress, and we evolve baby steps at a time. There is no behavior we could ever do that is not accepted into the arms of Unconditional Divine Love. Unconditional means just that, love without *any* conditions.

Hopelessness and Despair

You feel hopeless about life in this plane. So hopeless that you think about dying.

It crosses your mind that it would be easier to just not wake up one morning, because you can't take feeling this drained anymore. Drained from living around others and in a society whose values do not reflect your own. It drains you each time you listen to all of the negative news on television. It drains you to constantly be around all of the mean-spiritedness, disharmony, competition, greed, and violence that surround you.

It drains you to not feel supported for who you are. For your entire life, whenever you've tried to engage others in a conversation meaningful to you, you've heard, "I have no idea what you're talking about." Or "Wow, that's deep, dude," then they change the topic to something painfully superficial. But maybe not before they get in a few mocking jabs that knife you right in your heart. Your own family has never understood you. How many more times do you have to hear them say "you're weird" or call you "whacko"? Or "you must have been switched at birth" and "you're from another planet."

Feeling this hopeless about ever feeling supported, loved, and ever fitting in, you do think of suicide. You wish you could just end the pain. You know you're here for a reason, but you don't think you have energy left for the task. You don't really want to kill yourself, but you feel hopeless about this life changing for the better. You just want to feel love and to radiate love, and it seems impossible to live.

A client I had worked with regularly came in one morning saying she was a bit freaked out by some past-life memories that had spontaneously arisen since we had last met. In that lifetime, she was a man who was an infantry soldier at war, and the scene was of him and his male lover. They shared an incredibly intense love for one another.

"How much do our past lives influence us?" she wanted to know as she sat down.

"Why do you ask?" I suspected there was more behind the question than it might appear.

Essentially, she was realizing how she had always been looking for a love she never could find. She apparently had been reflecting quite a bit upon the promiscuous behaviors of her teens, the sadness she felt at not feeling very loved by her former boyfriends, and . . . upon whether or not she should try female relationships—maybe then she would feel loved.

She admitted to me that she felt so sad that she found herself thinking it might be easier to just not be living any more than to continue in this life feeling that little love. I got the impression that the love she felt in her past-life memory was so strong that it catapulted her into some spiritual depression and existential anxiety.

She assured me she wasn't suicidal, but she couldn't help thinking about being out of this body, out of this life.

"I have some bad news for you." I was thinking of how to tell her what I had learned from esoteric material and from near-death and past-life data. "I hear that whatever consciousness we are in at the moment of our death is the consciousness we are still in when we sever the connection with this body. Consciousness resides beyond the body, so killing the body doesn't work."

"So what I really need to do is resolve this in this life, so that I don't carry it over to any other lives . . ." She was clearly contemplating aloud.

"I guess so . . ." my voice trailed off, as I waited to hear where her mind went next.

"Are we just puppets?" She was clearly distressed by the idea. Her lips were turning blue.

"You want to know if you have free will?"

When she nodded, I told her that some do believe that we don't really have free will. Those are the ones who believe that consciousness is an epiphenomenon of the brain. I explained how they tend to think that our autopilot system makes us a victim of our brain. However, the physicists and consciousness researchers that I follow disagree.

I tried to figure out how I could describe it succinctly to her, so I referred to the Consciousness Map. I told her I absolutely believe that she has free will . . . as much as she does in any relationship. We both smiled at that, but she knew what I meant. I pointed out how, at the level of the first center, the neuroplastic body sometimes seems to have a mind of its own, regardless of what she may choose. But also how she could choose at the third center level what she does with her body and practice a "new neuroplastic normal" way of engaging the world to override that autopilot system.

With such mindfulness, she could definitely make decisions for her life, so much so that sometimes the soul interjects something different at the fourth chakra center level, and she could choose to ignore it. Yes, it's complicated at the highest levels, for there are many layers of self that want to have their say. It's like a large family where everyone has an opinion. It's great when we can collaborate, but sometimes one person gets their way more often than not. She may not get her first choice in every matter. But she definitely could decide with whom she would like to be in an intimate relationship.

With respect to her career, she could be a writer, make films, and influence the consciousness of the audience about topics like human rights, which are so important to her in this life.

Apparently, our lengthy conversation helped her, because when I saw her the next time, she seemed to barely remember this state when I asked about it. She reported that she had shifted her thinking and was no longer in that frame of mind. She then went on to ask about entirely different issues.

It's important that we all shift our spiritual anxieties and depressions. There is a world out there awaiting our spiritual light.

Initial Actions/Tools for Coping

What I would like to say to you is know that you are *not* alone on this plane. There are a growing number of awakened, conscious people here, maybe even 25 to 30 percent of us, according to a 2009 report from the Pew Research Center, and we need you. You have to practice holding steady your light so that it joins with ours and moves the world toward its next quantum leap in consciousness. Experiments suggest that it takes as little as the square root of 1 percent of us (called the extended Maharishi Effect) with coherent loving thoughts to usher in a quantum leap for the population. That's a really small percentage!

Plus, you have a unique mission that is your responsibility to bring in at this time. So we really do need you. Remember, *each of us as a soul is a unique energetic fingerprint or blueprint field of energy, here to bring our own particular combination and pattern of energies into the planet.*

And besides, I have bad news for you. As I told my client

mentioned earlier, if you were to kill yourself, your consciousness would be exactly the same afterward as it was before you offed yourself. Nothing would have been solved. You would be exactly the same. And since you have to work it out eventually, you might as well work it out while you're here. If you have to come back and start all over as an infant, you'll be even further behind.

- If you truly are suicidal, then please get professional help, preferably from a transpersonal practitioner who is a licensed mental health therapist.

- If you need medication to even get you to the point where you can work in therapy, please consider it, while you also learn ways to cope with the insufferable despair and sadness that you feel. Many of the exercises we've already covered can help you during these trying times.

- If you're this depressed, it is highly likely that you'll find your thoughts overly focused in an all-or-nothing polarizing way; on what you can't do rather than on what you can do; on who you don't resonate with, instead of even the few you do relate to. You'll also likely notice that you're assuming "no one" gets you or supports you, when perhaps you haven't really tried to reach out to the broader—even global— community.

- One of the first homework tasks I would also assign to you would be to really use the Internet to search for your keywords—your belief systems, your spiritual aspirations, your specific interests—so that you can find out who *across the globe* is also interested. When we just leave it up to those we meet in the neighborhood, or at school, or at bars, and so forth, then no wonder we don't see that there are many out there who would get us.

If you are interested in food as medicine, then search for those who are already teaching it. If you are interested in learning more about how energy relates to consciousness and healing of the body, then google it. If you are into conscious capitalism, then challenge yourself to find even one other on the planet who is into it. (Clue, five people a day in my office alone.) And if you don't see it at first, whatever the interest is, persist or check back later.

I've learned that different users receive different search results based upon their previous browsing history. Just as an example, as I was preparing this manuscript, I searched for "quantum psychology" and got back some totally unexpected findings, ones that were not there the last time I looked. So keep searching!

- Depression worsens when you try to ignore it. The earlier you go into it, to perceive the message it contains for you, the sooner it will pass. Try to notice when it is still in the state of melancholy or sadness, before it moves from a "2" or past a "5" on that 0 to 10 scale. To go into it, allow yourself to see the sadness energy as simply energy that moves, peaks, and dissipates. Watch it as an objective observer, without identifying with it, and notice its movement. As you do, you will notice the state shift into something else. Recently, a client allowed herself to do this, and was surprised that she shifted into Self, into Essence. In minutes. It is in *not feeling* it that the energy gets bigger and compounds, like a pressure cooker. So *feel*. And let the feeling *move you* in your life.

Grieving Loved Ones Who Have Passed

The more you touch and understand torsion fields and quantum consciousness, the more you feel confidence in its ability to keep you connected, even beyond death. However, just because you know that consciousness continues after it leaves the body, that doesn't mean you don't grieve the loss you feel in this reality.

Grief is a roller coaster ride, for sure.

You feel confused—happy for your loved one who has passed and is ready for their next experience but so sad for yourself and those of you left behind, who are really missing this very special person in your life.

Once they pass, you may feel some relief that they are no longer suffering and stuck in some sanitized institution. Or relief that a difficult situation is over. But then you feel guilty for feeling relief.

While they are still in the process of dying, you are hopeful one minute and in despair the next. Your insides feel shredded. Your stomach is knotted. Your mind leaps from place to place trying to find relief from the suffering. You try one coping tool after another in a desperate attempt to keep your heart from tearing apart.

Initial Actions/Tools for Coping

- Allow yourself time to cry, whether for five minutes or an hour. It releases the pent-up energy. But set a limit, then get up and do something nurturing for yourself.

- If you'd like, journal your emotions for ten minutes, allowing them to flow from your body and onto the page. Then get outside in nature and engage in activities that are uplifting and nourish your soul.

- Do some yoga, moving qigong, massage, or other physical activity to move the grief energy out of your body.

- Try meditation imagery as a variation on prayer and intention. If you'd like, envision Christ, the archangels, saints, or some other spiritual entities as being in the hospital with your loved one. See their energy as sparkling light, moving through your loved one and any surgical staff, signifying the process of healing for the highest good, whether physically in this realm or spiritually as part of their passing.

- As far as self-talk, be your own spiritual counselor. Remind yourself that experiments suggest that your loved one's brain is likely picking up these prayer energy signals, even if your loved one is consciously unaware of it. Find the thoughts that feel affirming for you and practice them in your head like mantras.

When my ninety-year-old father suddenly became ill and weak in what turned out to be the last month of his life, I thought to myself: "It is entirely possible that God works a miracle through my father. Dad's soul can heal his body if it is in its highest interest to do so. Yet I have to accept that when Dad passes, it is because his soul is ready for a new vehicle, a new expression, a new life form, and that now his body is getting so worn out that much of what gave him joy is lost to him. He even had to give up golf this year and can't make the drive to the horse track. So I need to be understanding and unselfish in my prayers for him."

I also practiced gratitude self-talk, being thankful for what we had shared, the love we'd been able to express over the later years, and how hard he'd worked over the decades to provide for his family, even though he hated his job. I expressed thanks for having parents who loved one another even past their fiftieth wedding anniversary.

• Forgive and let go.

Chapter Five

Developing New Definitions of Love from Transpersonal/ Spiritual Experiences

During your spiritual or transpersonal experience, you likely have experienced a consciousness state of incomparable unity, connection, and wholeness . . . the ultimate Love. Returning to your everyday life, you are painfully reminded that this material world appears to lack the unconditional love you keenly remember. Being quite human, your response to this situation is to find it difficult to maintain your unconditional love for others. Then, out of self-judgment, you lose your unconditional love for yourself as well.

In this chapter, we'll take a look at some of the common pitfalls those who are coping with transpersonal experiences can fall into where love is concerned, including how to maintain unconditional love for yourself and others.

Perfectionistic Standards and Expectations

Your experience of wholeness gave you a vision of perfection. Now you believe that it's all up to you to fulfill it. The responsibility feels overwhelming.

You strive to make everything perfect around you. You want and expect the very best from yourself and others all the time and in all scenarios. You project your ideals onto everyone else and expect that they should live up to your standards. You are overly focused on all the things you think you must improve or correct. You're constantly focusing on mistakes and missteps made by your spouse and children. There is always more to do, more to improve upon, more to fix, especially within yourself and those closest to you.

You have to be the perfect parent. The perfect partner. The perfect worker. The perfect breadwinner. The perfect housekeeper. The perfect host or hostess.

Some of you believe you will be judged by God at the pearly gates for how perfectly you have lived your earthly life. And you are afraid of falling short in his eyes and not being allowed into heaven. (Though quantum physicist, chemist, and neuroscience researcher William J. Bray assures us, after his more than thirty medically documented "deaths," that we are loved purely and unconditionally, and we are definitely not judged by God.)

You find yourself constantly oozing anxiety all over the place. You actually catch yourself believing that if others feel your anxiety, then maybe they'll hurry up, or do it better, or try harder, or listen to you more, or take you more seriously when you ask for things to be a certain way.

Feeling blocked by others and kept from achieving the perfection you work so hard for in each and every moment, you find yourself feeling anger. Then judge yourself for feeling anger. You say to yourself, "It isn't spiritual to be angry." So you shove it down deep inside. The resentment simmers, always there on the back burner. It doesn't take much for it to flair up.

The intense scrutiny and hypervigilance drains your energy. The magnitude of it all weighs you down. How exhausting! You're tired of being tired.

Others complain at how controlling, intense, or intimidating you are and how nothing they do is ever good enough for you.

Yet you are stuck in this habit of judging things as not perfect enough day in and day out and have no idea how to think any other way. You even judge yourself for judging.

Some of you have even begun to hate yourself for it and for all of your "clearly obvious" imperfections, which are all you tend to see, since that is all you tend to look for in yourself. Psychology has a name for this habit. It is called perceptual or inattentional blindness, where you are blind to everything except the one thing for which you attend or filter. No wonder you can't see the good, when you continually look for the bad that has to be improved upon. The glass really is always half empty in your eyes.

This all-or-nothing, always-or-never viewpoint, this pendulum swing in perception, *from either it's perfect or it's a catastrophe, it's good or it's bad, it's holy or it's evil, it's total success or it's utter failure,* is exhausting and doomed to keep you in perpetual anxiety, frustration, disappointment, and resentment. No wonder you feel mood swings. This entire line of thinking is a no-win scenario, and it keeps you anxiously trying to beat those odds, until one day you collapse under the mindset of hopelessness that is depression.

I've been through it and know it all too well. Maybe not quite to this extreme. But I do remember what it's like to envision perfection with my sixth and seventh centers, to strive for it in work and life, and to feel anxious and resentful all the while it remained elusively out of reach, until I developed the chronic lung issues that led to my changing out of my very Type A way of life. I very much recall how others would let me know how intimidating I was to be around and how intense my energy felt to them. Not exactly the loving energy I was going for.

I remember, too, feeling utter confusion about the entire concept of judgment. I wondered how it was even possible to not judge, seeing judgment as a byproduct of the intellect and believing it to be inevitable.

Perfectionism can be paralyzing. How many of us grew up with the old saying, "If you can't do something right, don't do it at all"? So we don't. One young man who came into my office felt paralyzed in this way. He even called himself a perfectionist right off the bat. He was very aware of the problem that caved him in but clueless as to how to find a way out. He liked much of what he did for a living but resented the unlivable wage he was paid at his nonprofit social services job. His belief that only nonprofits could possibly be "conscious" places to work kept him from even considering pursuing higher paying positions within any corporation.

Because of that mindset, he could only see a life of financial insecurity in his future, a thought that drained him of the energy and motivation to do anything different in his career. Depressed, he would resort to wondering about going back to school for a master's. But he would talk himself out of it, because he could not be certain that he would thoroughly enjoy it nor that it would lead to the perfect career and life situation anyway, so "why bother?"

Referring again to our Consciousness Map, we can size up perfectionism this way. Perfection is glimpsed as we move through our seventh center, even subconsciously or not consciously in this lifetime. The sixth gives us insight as to how we can bring it to fruition in the first, the material realm. The fifth urges us to carry it out, and the fourth provides the loving responsibility that motivates us through the task.

However, when perfectionism causes us problems, we could say that we are being an all-or-nothing ego thinker in the third center, which likes the power of achieving our goal alone. It seeks to compete and control to keep the upper hand on the project of life, but as it does, anxiety continues to mount in the second until it overwhelms the first center body and exhausts us into a state of second center depression.

A woman in her forties told me she had fought depression for as long as she could remember but was in a particularly low spot since she had recently lost her position at work. Once the head instructor, she had been fired from the studio where she had learned to dance as a child. She felt devastated and ashamed of what others would think of her. Her career achievements were integral to her sense of self.

As we talked, it became clear she was driven by perfectionism and came off as very judgmental, intense, and intimidating to others. She said she had tried Western psychotherapy and did not want to take antidepressants. In my office, she explained the hope she held that a more spiritual perspective would work for her, though she didn't consider herself religious and wasn't quite sure what "spiritual" really was.

The day we discussed all emotion as energy, and each as a "same state consciousness," was a significant breakthrough for her. She came to understand how, within a state, a certain logic prevails. But outside it, a different logic prevails. Each state of energy, emotion, or consciousness has its own way of perceiving and thinking that doesn't necessarily translate to other states. That is why we can journal about our depression one day, then a few weeks later from a happy state, reread our journal and not even relate to our thinking within the depressed state.

She came to learn which thoughts, environments, colors, situations, imageries, scents, tastes, and sounds were associated with her depressed state of energy and how her body reacted physically when she was pulled into that state. She seemed to get it that these arise together, much like a constellation of stars appearing together; when she enters the constellation through one trigger pathway, it is like all the other "stars" come into view, blinding her field of perception. Specifically, when I had her construct her *spiritual, or "unconditional love," constellation,* it opened up a completely new set of possible ways she could focus her mind, which could be very healing for her and for her relationships.

She realized that lilac was a scent that immediately calmed her and took her into this feel-great constellation; that the outdoors and nature were something she needed to access more (either in her physical life or in her imagery), especially a particular beach scene; that there were particular thoughts she *could focus on to stay in and even amplify this state of love*; that she could become mindful of her breathing or her heart beat, that she could slow them purposefully, that she could send her heart and lungs love and gratitude, and they would fall naturally into the rhythm of this love constellation. As it turned out, this exercise and its homework worked very well for her.

The next session, we took her imageries deeper. I asked her if she would be ready to open herself up completely, so that the Universe could speak with her, either directly through her soul, or through a guide, or through some part of her physical body, or symbolically. A smile flashed over her face. By the end of the session, she looked at me as if stunned, and uttered, "Amazing!" What she had seen, she said resonated with her at her very core.

One of the scenes that came to her was of herself as a thirty-something-year-old woman in the 1800s, living on a very dry and rugged land and fighting to keep food on the table. She had several children, whom she loved, but she was miserable. My client just knew that her former self had killed herself in that lifetime. As she scanned several points in time, my client realized that she had been born in Scotland, but at fourteen years old, her parents, who could no longer support her, had married her off to a man she did not love, who then forced her to come to America and leave behind the green and fertile land of her happy childhood.

As my client reflected upon that lifetime, she seemed to sense that *she had brought that depression consciousness with her into this lifetime,* as well as her inner conflict about motherhood and her *extreme avoidance of poverty through a perfection-driven focus on material success.* As we talked, she appeared eager to leave that depression state of consciousness behind her.

She also spoke to her future self, an eighty-year-old woman she encountered on her front porch swing, holding hands with her husband of over fifty years, and gazing out at the yard where she could see her five grandchildren at play. From this facet of self, she was told that she needed to quit hating herself and to stop chasing achievements that would never satisfy her soul; that it was her relationships that had made her the most happy over her lifetime, both personally and in her career; and that she needed to stop pushing people away and be more gentle and compassionate with them.

The following week, I asked her about how meeting up with past and future selves had impacted her in her daily life. She said it *had completely altered her worldview.* Though she had never been religious, she now felt that she did have a soul or an aspect of herself that was eternal. It made her really motivated to learn whatever lessons she needed to learn so as to not have to keep reliving the same types of depressive lives. It also validated a feeling she'd had since childhood, a sense that she came into this life with her depression and that it was not something that stemmed from anything particular in this lifetime.

She also was now regularly meditating on "unconditional love," which she associated with gold energy, and was actively sending gold love energy to anyone in whom she perceived tension, as well as to herself. (I had told her about some research in which it was found that the subject's brain registered someone sending them a telepathic message, even though they didn't register it consciously.)

She shared some unusual interactions she'd had that week as a result. Smiling, she gave some examples of when colleagues had delegated some activities to her that they usually kept for themselves; when a female dancer who usually snubbed her seemed more friendly; and when she no longer kept to herself, making assumptions about people, and instead, reached out to others. She said that where normally she would have expected an energy-draining experience, reaching out turned into quite a positive one. She found herself feeling warmer toward others, more compassionate, and as a result her mood was shifting into a much lighter place.

And that was just after one week of practice.

Initial Actions/Tools for Coping

What helped me the most was the insight that, on this plane, *perfection is achieved collectively, not individually.* We are intended to collaborate, to bring our own unique flavors into the soup of creation. In material reality, *perfection lies in our diversity,* the diversity that is group and collective consciousness. Our responsibility is to do our individual part and calling. Focus on *what* you can do, *when* you can do it.

Remember, evolution of the material realm is s-l-o-w, because the physical is working through electromagnetic energy, the slow-moving kind of energy. Baby steps are how our world evolves gradually over time, despite the occasional quantum leaps. You are not excluded from this physical law of God. Give yourself a reality check . . . the human body is a physical and material thing that can break. If you really plugged into the torsion plus fields that truly are closer to the power that is God, you would explode immediately.

So you have to quit competing with God for the prize in perfection.

- Schedule into your week some goals you would like to work toward, then schedule in some relaxation and fun to balance your physical body. Over time, you will see your progress, and you will feel good enough to enjoy the fruits of your labor. That is how to focus on what you can do, when you can do it.

- Regarding your thoughts, *allow discernment to replace judgment.* In *judgment,* we categorize or label things as positive or negative, as if we have to file them away in some judgment filing cabinet equivalent of heaven or hell. Whereas, in *discernment,* we may notice and observe qualities of consciousness and character, but we do not judge. Much like with snorkeling, when we are amazed at the diversity of nature, the corals, the various fish, and we're simply enjoying the bounty of sea life. We don't need to judge which fish we like best.

- To shift out of the paralyzing pattern of perfectionism, we need to become more aware of our all-or-nothing beliefs and expectations. Notice the patterns or themes of your thinking. Observe where and how often you categorize things in black-or-white terms. Practice noticing what others and you *are* doing that is helping to achieve those larger goals of life.

Expectations of Perfect Love

You find yourself feeling tragically brokenhearted, and sometimes you aren't even sure why. You burst into tears because you can contain the energy no longer.

A melancholy hovers over your relationships. You try so hard with your mom and dad, with your brothers and sisters, with your friends and neighbors, with your children and intimate partner. Yet it seems as if no one ever returns that kind of full and constant love to you. Or what they do never quite feels enough.

You exhaustingly feel as if you are always the one who gives, who nurtures, who calls, who reaches out to care for them, who attends to their needs.

You long to feel that kind of love. The kind of love that scoops you up in its arms and holds you tight. The one that simultaneously leaves your body limp with trust and sizzling with orgasmic joy. The kind that listens intently and takes to heart every word you say. The kind that remembers, even months later, your conversation about big dreams and small pleasures. The kind that surprises you without reason or occasion with your favorite flowers or perfume, tickets to that Broadway play or football game, a weekend getaway to your "secret" fishing hole, or vacation plans to that spot you've always fantasized about visiting. The love that wants for you what you want for yourself and considers you an equal priority in life's imaginings.

You routinely try to ask for what you want, thinking it best to communicate your needs and feelings. But it never seems to do much good or for very long; the ghosts of talks past haunt your days and enter your nightmares. So you try again, and again, falling more into secreted despair each time you feel you have to revisit the topic but return empty hearted.

The disappointment and resentment build, but you keep shoving them down. You think you have let it all go, yet the pain simmers there underneath the cover of forgiveness until some sort of major loss jerks away the sheets of blame and exposes you to yourself.

Your neck and shoulders ache with the weight of holding you up. Your back stores the pain that your heart can't bear.

Several women and men have shared with me their belief that the perfect and ecstatic love they experienced in their transpersonal states of wholeness was possible to feel with a spouse. So they searched, and from relationship to relationship they moved, trying to find perfect love with another human. Until they were so frustrated that they came to talk in session about their hopelessness of ever finding love.

For one man in his fifties, the realization that he had been futilely seeking perfect love his entire adult life happened in one of our sessions.

When we had first met, he had complained of feeling depressed his entire life. But it was currently at a peak, for he had just gone through a divorce. As our work progressed, I began to sense his higher consciousness tucked underneath his overwhelming feeling of responsibility, the suppression of his deeply creative self, yearning to be set free, and his perfectionist expectations for himself and others that had kept him imprisoned in his "boring though secure and responsible" life as a father and husband.

This day, despite having earlier identified himself as an atheist, he agreed to let me take him deep within his being.

I hoped he could have a talk with his higher quantum self to hear what he needed to hear to get himself unstuck and to move forward with his conscious creativity. But the higher self always knows the root of what is going on and will show us exactly what we need to learn next. In his case, it was about his expectations for feeling perfect love from his human relationships and the resulting chronic disappointment and depression he had felt his entire life.

As I led him into subtle awareness, he began to calm, then to quiet, then to listen. Really listen to his inner self. He followed the imagery he was receiving into a blue star cluster that he thought might be some kind of galaxy.

Once there, he encountered gentle beings in their energy bodies. One energy being in particular he felt drawn to and wondered whether it was an aspect of himself or some other being he was close to, there at the "home" he was originally from. On one hand, it felt like a father figure to him, much older than him and radiating wisdom. But it also felt to him that he was face to face with his Source or higher self.

Upon feeling engulfed with the love that radiated from these beings and this dimension, he began to get choked up with emotion. Soon tears were streaming down his cheeks. When he ended his meditative state, he held his head in his hands for a long while and just sobbed. He wanted so badly to feel that kind of love here on Earth. He realized that he had felt depression much of his life, because he could never feel love the way he thought it should feel. Now he understood why.

Near the end of our session, we had to process how that kind of love from Source can't quite be felt in the same way in this physical dimension. And how he couldn't expect any human (familial or spousal) love to live up to that experience. But that he could learn to find relationships of higher consciousness that could, at least, begin to approach the kind of unconditional love for which he had always yearned.

Furthermore, we used the Consciousness Map to help him differentiate the various kinds of love he might experience, depending on which state he found himself at any given time. And we used the language of quantum physics to help his disbelieving logical mind understand the experience he had just had; how it might be that he could be in more than one place at the same time; how he is a being that is multidimensional and holographic in its consciousness, manifesting here in time and space but existing nonlocally as well, beyond both time and space as we define them.

Initial Actions/Tools for Coping

- Notice the similarities between believing in perfect love here on Earth and the perfectionism section above. Go back and read some of those suggestions. Address your all-or-nothing, unconstructive self-talk.

- Remind yourself of the ways your family, friends, and partner do show their love, however imperfectly. For help, read *The 5 Love Languages* by Gary Chapman. And to his five languages, I would add a sixth, the felt sense of the energy of love. In addition to the other ways you feel love, most of you feel subtle energy and therefore can feel the energy of love beaming at you and can feel when it is absent.

Consider that if your loved one has sealed off his or her heart because of their fear of feeling the vulnerability of emotion, that doesn't necessarily mean they don't love you. It may simply mean they have numbed to emotion and in so doing they have closed off their heart. To you it will feel like "nothing." Like there is nothing radiating from their heart. And it may be essentially so because of the way energy follows thought. Yet they may love you intellectually or physically, and you may have to decide if that is enough.

- Use the Consciousness Map to realistically think about the kind of love you can feel in each of your centers. Give yourself a reality check about the kind of love your partner is capable of. For instance, in center one (the physical), there are physical expressions of love, such as sexual union, hand holding, sitting closely together on the couch, cuddling in bed, giving presents to one another, doing fun activities together, and doing things to help one another.

In center two, you feel love as the energy of emotion, of empathy, and of the caring that drives you into relationships and into the desire to start a family and have children. In center three, love is experienced as listening and being considerate, as thoughtfulness, mindfulness, mental harmlessness, respect, and patience. Center four is where love unfolds its more "we" or collective mindset and unconditional quality and is expressed through unconditional kindness, compassion, forgiveness, responsibility, reciprocity, and collaboration.

Center five love is through creative expression, where love moves us to create for the betterment of the "we" and in so doing, to evolve our worldly experience. Through center six, love comes as a vision for how "we" could be, for how our creative expression might look. And through center seven, love unfolds as the passion of inspiration that moves creative expression and as the higher sense of "knowing," realization, illumination, purpose, wisdom, and sacrifice for the good of all.

Enmeshment

Enmeshment is when your relationships become so intertwined that healthy individuality is lost. "We" thinking is such your natural state that you forget what the "I" stands for.

You do everything together with your partner. Even go room to room with him or her. And you believe this is the most loving thing you can do.

You continually sacrifice your wants and needs for others' desires and preferences, to the point where dysfunction is occurring in your life.

When someone asks you, "What would you like to do Saturday night?" You automatically reply, "What would *you* like to do?" It is such your habit that you may not even know what you would want to do if it had to be your choice.

For you, it's all about maintaining the peace and love among you as a group, but to the degree to which you or others in the group have lost the sense of self and the purpose of individual creation.

One twenty-year-old woman vented a great deal to me about her frustrations. Her fifty-year-old father kept "guilting" her and demanding that she pay all of his bills, while Dad preferred to go play and not work. Also, this client's brothers and cousins kept moving in and out of her home, bringing their girlfriends or wives along, and not even thinking to ask if that was okay. They just expected it of their sister/cousin because she was doing pretty well, in their eyes, and making a decent living (for the first time in her adult life).

She told me that she thought that's what loving people did—whatever anyone/family expected of them—and didn't know how to set her boundaries or say "no" in a loving and fair way. As we began to talk, she started to realize why she couldn't build momentum for living her soul life and why she regularly felt so overwhelmed and chaotic. And how much she needed to learn how to set healthy energetic boundaries and to leave behind the enmeshed mindset in which she had grown up.

With help, she set healthy boundaries, including setting an end month date for paying Dad's bills. And, as an example of the results, she told me later that her father got a new job, making a large salary, "just in time."

Initial Actions/Tools for Coping

In enmeshed relationships, we sometimes have thought that it was spiritual to rescue our loved ones from their harmful or dysfunctional ways, and we have set up our life to do that. Understanding the difference between helping others out of love and rescuing them to the point where your life revolves around them in an unhealthy way is essential to setting good boundaries.

Helping is teaching or moving someone toward healthy independence in a balanced manner. *Rescuing* (in this context) is doing for someone what would be healthier for them to do themselves. Sometimes it is done as a relationship pattern because the rescuer's ego gets some psychological gain out of it and/or because the one being rescued keeps overly demanding it, and the rescuer doesn't set a boundary. Usually rescuers marry people who enjoy being rescued, but then wonder why they begin to resent the spouse with whom they have this energetic cord.

- Consider that healthy loving relationships are built upon the foundation of two whole "I"s—largely independent healthy individuals—each person taking responsibility to ensure their own sustenance, happiness, and fulfillment. As they do so, each radiates that happiness, that wholeness energy, into their relationship. From such a state of love, the relationship more naturally feels *allowing, mutually supportive, and reciprocal.*

- As you might, don't give independence a bad rap and see it as an all-or-nothing selfish enterprise. Independence means we can take care of our own needs and desires, *and* we can ask for help as needed and appropriate along the way, *and* we can each collaborate so that the relationship feels "two-way" and not "one-way," where one person is doing most of the work to maintain the relationship.

- For those of you who come from a tightly enmeshed family upbringing, where you find it difficult to be true to your own soul, separate out your thoughts from the thoughts of your parents. Take any morsels of wisdom you find from them, but seek out your own wisdom that pours forth through your higher centers of consciousness.

- If your parents have raised you with an enmeshed mindset, they may have expected that you help them in any way they ask it of you. That you be together for most every family event and gathering. But *none of us is capable of all of that* in our adult life.

Set realistic boundaries. Know what you can do, and what would unbalance the scales of your own well-being. Take responsibility for maintaining your balance in a very preventative way, because life does toss us curve balls, and we have to save some energy to handle them. Or, mixing metaphors, as much as possible, try not to let your energy tank drop to less than 50 percent full as a regular rule of daily life.

- If you are a rescuer, stop waiting for your partner to become less needy of you before you can make changes. That is not likely to happen if you have a strongly established pattern or energy cord between you. Stop blaming them; take responsibility for helping rather than rescuing and for setting your own boundaries. Learn to say "no" in a way that is appropriately firm, yet respectful and kind.

Be practical and don't pendulum swing in an all-or-nothing style into the opposite direction. Instead, use common sense about what and how quickly your partner can take on more of what you were doing for them all this time. Communicate and collaborate toward a healthier relationship with much patience and compassion.

Difficulty Receiving and Loving the Self

Love yourself? You can barely think the words, much less act on them. "That's selfish," you believe.

You think, "I'm supposed to love others, to sacrifice my needs for theirs. That is the most important reason I am on this planet."

But now you feel drained, without energy, and depressed.

You can barely remember what it feels like to feel connected to your soul.

Mistaking self-love for selfishness, you have allowed yourself to give until you are empty. You have lost your sense of self. You may not even know who you are anymore.

Your intentions may have been loving, but the effects have been disastrous on your health and relationships. You may have developed chronic health conditions. You may find yourself resenting your family and friends so much that you fantasize about moving away and starting over. Sometimes you even wish for God to take you home one night so that you don't have to wake up and struggle to bear another day of this life.

Initial Actions/Tools for Coping

- Differentiate self-love from selfishness. *Self-love* is when you allow yourself the conditions for your soul to thrive. Like a plant, which needs a certain amount of water and sunlight and a specific type of soil and nutrition in order to bloom. It is not selfish for you as a flower to allow yourself the conditions for which you were created. On the other hand, *selfishness* is an out-of-balance state. You are being selfish when you disproportionately take, constantly demand that others revolve around you, and hoard more than your fair share of time, attention, or resources.

   Balance your needs with the needs of others. As far as your soul work, give others of your time and attention, create in the fifth center what you came here to create. Do what you can, when you can. Then relax and nourish the physical body by getting plenty of sleep, movement, water, and nutrition. Destress, find your peace, and play. It feeds the soul by putting you in a state that is conducive to perceiving interconnectedness, downloading intuition, and setting off the synchronicities characteristic of the soul's Law of Attraction.

- Balance *yang* with *yin* energy, the outpouring with the inpouring. Give. Then receive and feel gratitude. Give. Then receive and feel gratitude. Repeat. Repeat. Repeat.

- Nurture and love *the quantum self,* in all of its aspects, balancing each of the centers—*physical, emotional, mental, and spiritual.* As you learn to love the self and allow yourself to flower, you will find yourself filled with the energy of love. It is then that you will radiate love as energy out toward others. In so doing, you change the world.

As a specific exercise, journal about who you think you are, physically, emotionally, mentally/intellectually, and spiritually (PEMS). In each of these four areas, write down qualities, characteristics, traits, skills, strengths, and know-how. Make sure to notice what energizes you and brings you joy (i.e., inpouring energy); write down those topics, interests, people, and activities. Make note of what inspires you toward your highest creativity (i.e., outpouring energy) and which seems to point toward career direction. List which "outpouring" activities nurture "inpouring" higher states of spiritual attunement.

As you come to know more about who you are in this plane of material activity, take note of which aspects of this PEMS self you are *asserting* in your life. And which you are not. Note which aspects of self you minimize or feel resistance toward.

For example, maybe you give yourself credit for being a "big picture" thinker, but dock yourself for not being a "detail" person (there's that perfectionism habit again, thinking you're supposed to be all). Consider that *qualities* of self, and quantum consciousness at large, are simply energy. And energy may be like a two (plus!) sided coin; in this instance, "big picture" vision on one side and "detail" vision on the other. Focusing on the "positive" translates to *focusing on the side that is your quality to manifest,* without expecting yourself to have it all.

- To regain your sense of soul self, meditate on *the subtle energy within* you. For your meditation, practice "listening." Start by listening for sounds, let's say, in nature, such as a bird chirping. Bring full awareness to whatever it is you attend. Notice that when you do, the mind is open and receptive. That openness is the receiving state that is prerequisite to hearing the small voice of your intuition, or the voice of God, or perceiving subtle energy in the body, or getting your creative downloads. Which is likely the full point of meditation in the first place!

Listening is the quickest meditation I know of that empties the mind. When you listen, truly listen, you cease all mental activity in favor of receiving that which is beyond your normal limited self. So now listen and receive the voice of your highest self.

- Try the ancient *"Who Am I?" meditation.* Over and over, allow this question to arise in your mind, and notice what answers bubble up. Write them down if you'd like. Don't stop. When your mind goes blank, wait for another layer of answers to bubble up. Then when your mind goes blank again, wait for yet another layer of answers to arise. Repeat. Until you just feel like you have to stop.

I warn my clients ahead of time that they will undoubtedly feel like smacking me before we actually stop. That seems to prepare them to cope with the discomfort of peeling back the layers of identification that have been rigidly stuck there for maybe a lifetime or so. In this spiritual exercise, you will most assuredly notice that the first few waves of answers are perceptions of identity that are associated with your lower centers of consciousness. Only after you exhaust and dismiss those perceptions of identity as who you are will your upper centers begin to be heard in your inner awareness.

• And remember, *you are not any one of those things; yet you are a unique stream of energy coming into form as those things, emerging from the field of quantum potentiality that some call quantum consciousness or God.*

Envisioning Healthy and Unconditionally Loving Relationships

As a higher consciousness person, you may be wondering, "What does a healthy and unconditionally loving relationship look like, anyway?" You long to surround yourself by such relationships but haven't really seen any to model yours after.

Consider this model.

Imagine you have learned who you are, physically, emotionally, mentally, and spiritually. There you are, taking responsibility for living your authentic higher self . . . that stream of energy, those qualities that are at your core. And as such, you are feeling largely happy and complete.

Being who you are in your life, you attract those who cherish you for exactly who you are. And that feels amazing! It feels freeing, nourishing, and expansive.

Being healthy, you tend to attract healthy partners. Ones who know how to make themselves relatively whole, how to give and receive in balance, how to share, how to collaborate, how to make your dreams as important a priority as theirs, how to take responsibility for their own actions and emotions, how to forgive themselves and others and move forward day by day.

So there you are, living your authentic life, feeling pretty good on a day-to-day basis. You are filled with love and joy, so you radiate that into your relationships. A loving person, you are naturally loving toward others, in a way that is devoid of the conditions normally found in ego-based relationships. They can be who they are. You can be who you are. Life is good. You support them for who they are. They support you for who you are. The relationship feels natural, free, loving, unconditional, and sustainable over time and growth.

This is a relationship that feels mutual, reciprocal in its energy flow, and enlivening due to the energy of love radiating between you.

Neither of you will ever be perfect in your interactions, but with compassion, forgiveness, and responsibility in the air, love endures.

Initial Actions/Tools for Coping

- Do the PEMS exercise for yourself. Write down all you know about yourself physically, emotionally, mentally/intellectually, and spiritually. Note how much of you is being asserted in your life or how much of you feels passive and needy. Take responsibility for your own fullness and joy. Get help if needed.

- Do the PEMS exercise about your partner. Write down all that you have learned about your partner and what makes them feel joyful, their dreams and aspirations. Assess how much you have truly listened to your partner or how much you have made the relationship about you. Ask how you can respect and support your partner in his or her life dreams.

- "How I Want to Feel in My Relationships" exercise: Write down all of the keywords that arise as to how you want to feel in your relationships. (If you would like, you can place each keyword in the PEMS section it most fits . . . P, E, M, or S.) Then, when you are asking yourself about a particular relationship, notice if you feel in that relationship how you really want to feel. If not, it may be time to let it go and wait for one that does feel largely healthy, adventurous, peaceful, supportive, intellectually stimulating, freely creative, or however you prefer to feel in your significant relationships.

- Instead of compromise in the usual sense, *take turns getting first choices* in daily life. Take turns picking the movie, restaurant, vacation, expensive toys, and even investing in one another's dreams. Don't be anal about this, but aim for reciprocity over the long run.

- Remember that none of us is perfect, so you aren't waiting for the perfect person. You are searching for the relationship that feels healthy enough, loving enough, reciprocal enough for you. Forgive yourself and your loved ones.

Difficulty Coping with Unloving People

You just want to love and to feel loved. Yet you are surrounded by people who make it extremely difficult. For they are hurting one another and you. Living out of fear, they are anxious, impatient, grabby, unkind, disrespectful of others and the planet at large. Many are angry, mean, and outright abusive. Some are violent and destructive, both locally and on a larger scale.

It is a scenario that makes it difficult to love them. Yet it is the mark of a higher consciousness (that is, faster vibrational and more inclusively loving) person to love . . . without exception.

Here's a story for you. I had just discovered qigong. In this particular month, I had attended my first, second, and third weekend of qigong training. Three of four weekends of qigong tends to put one in a pretty high feel-good state. Let's just say I was beaming with love energy.

It's a Monday, and I'm heading into the on-campus room where I regularly teach a life skills class to teens in a residential treatment home. I see two rival gang members out of their seats, standing face to face, fists raised, and ready to fight. All eyes are upon them. Two homeroom teachers are paralyzed in the corner. Without thought but responding instantly and intuitively, I walk up to the two boys, and, in my current and natural state of joyful love, I smile and practically sing out a "Hi Marc! Hi Eddie!" And *I beam. Literally, the love I am carrying beams* all over them. I can feel it emanating from my heart center outward.

The closest to me, Marc, swings his body around, reacting to being approached. He towers over me by a foot and a half, as does Eddie. There is a three-second silence as Marc looks me in the eyes, his fist still raised above his head. In the next second, he drops his arm, spins around toward his seat, and says in surrender, "I can't be angry 'round you!"

He went and sat down, as did Eddie. I gave none of what had been happening any energy. Instead, I directed my focus to beginning class. (But I couldn't help catching a glimpse of those teachers looking at me, apparently stunned.) Class went well and without any further incident.

Now let's face it. What happened that afternoon defies conventional classroom-management logic. But it was the most effective interface with "my kids" that I have ever experienced.

In quantum physics lingo, I'd say that I was carrying a *highly coherent energy field of love* that shattered the albeit coherent set of anger energy of the boys in such a way that their energy field shifted in a literal instant. And around such love, they could not hold onto the anger energy, because it was dissipated by the power of love. So they sat down.

I learned an incredible lesson about the power of unconditional love that day. Try it using the steps below as a guide.

Initial Actions/Tools for Coping

- From within the light of a compassionate mind, send them, those most difficult for you to love, *love as an energy.* There is no one way to do this. You can see love energy as a color, perhaps green, pink, white, or gold. Pick whatever color love energy is for you. You can see that color extend from your heart center to theirs. You can see them surrounded by that color or combination of colors.

- Think of which is your key "loving activity" out in the world. Direct it toward all others as inclusively as you can. Think of this as your *life meditation.* When you "wander" from it, just as with thoughts that wander in a sitting meditation practice, get back at it. When you wander again, get back to it again.

- *Do what you can, when you can.* Yes, that is a theme. Love in the way you can, as you can, toward yourself then others. Be patient and forgiving of yourself, so that you are in a state conducive for being patient and forgiving of others.

- Find your *compassion* and use it, regularly. Compassion is a quality of the fourth center, as is the sense of loving responsibility and forgiveness, which are included within any act of compassion. An aspect of wisdom, compassion is a mix of third center mental and fourth center loving energy. It is a loving understanding, a wise kind of love.

If you need help finding your compassion in a given situation or toward a specific person or group, then ask yourself what sort of conditions exist that may have created or led to the particular situation or behavior in this person. Consider that very few people grow up in unconditionally loving and healthy environments. So it is hard for them to be loving if they have yet to feel the energy of love in their lives. Harder yet when they are surrounded by hateful and vengeful energy. Hate begets hate, as love begets love.

For extra credit in compassionate actions, join any of the global peace initiatives and intention experiments that aim to reduce violence around the world. As was pointed out earlier, it reportedly takes as little as the square root of 1 percent of the population to effect change on the population when higher consciousness is involved. Get involved!

Chapter Six

Finding Happiness, Meaning, and Purpose

An Internal Shift in Values, Meaning, and Purpose—Even Happiness

What once held value, meaning, and purpose for you no longer feels as moving. You feel confused about what to do with your life. Should I change careers? Does this relationship match me? What do I want most out of life? What is my purpose? What is the meaning of life?

Or what brings purpose, and therefore meaning, on some days suddenly doesn't hold your interest on other days.

The reason is that *when you shift the center of consciousness from where you used to predominantly live into another, values shift. Meaning and purpose shift. Happiness shifts. Power shifts. Love shifts.* In essence, you can tell where you are within the centers in any given time period by the values you hold, by the thing or no-thing that brings into your field a perception and felt sense of meaning and purpose, and by that which seems to bring you a state of happiness.

It feels less confusing once we learn to "hold steady our light," but in our early jumps into collective or transpersonal consciousness, we move in and out of many of the various centers, oftentimes all in the same day.

For example, in *center one,* we value the physical and material, so focusing on accumulating material belongings that help us feel secure is what brings us a personal sense of purpose and therefore meaning. In *center two,* we value relationships and the feelings we get from the experiences of life and human love. Therefore, those take on personal meaning and purpose more than gathering belongings. In *center three,* we come to value mental activity, learning, analyzing, problem-solving, and achieving in these kinds of ways. Our purpose in life now feels like those kinds of intellectual and vocational pursuits, personal empowerment activities, and personal achievements; so those are the tasks we focus on, and those are what bring us meaning and personal satisfaction.

By the time we shift into *center four,* the lower personal senses have become increasingly aligned with the higher collective energies. Meaning and purpose are now all about higher and inclusive love. Love as the energy that coheres us all together as one. Love as the energy that comes in through our heart and head and unifies all of the centers. We value what is good for the advancement of the collective group of us, whether our community, state, nation, planet, or cosmos. What brings us personal purpose and meaning is the responsibility that we can take on for that higher good and the process of collaborating with others to achieve that higher goal.

In *center five,* we gain a sense of personal purpose, and therefore meaning, when we actively do our part for the collective. In *center six,* tuning into the higher vision and working with others to make it a reality is our priority. And in *center seven,* we are so focused on that which interconnects us with All That Is that we lose awareness of personal importance altogether. To others, it appears that sacrifice is our higher purpose and that which brings us meaning; to us, however, it does not feel like sacrifice, because there is no sense of loss. Only wholeness. We are in a state of inspiration, where we breathe in Spirit and breathe out love. Divine Will is our will.

And notice, *when our life actions are aligned* with the highest frequency values we perceive through the quantum self, it is then that *we will feel our greatest happiness.*

So what this means for you is that you have to feel inspiration entering through your seventh center, see it in the higher mind through your sixth, decide how you want to put it out into the world through your fifth, and choose those to whom you will bring this higher form of love through your higher heart, the fourth center. Then, use your lower personal mind at the third center to take care of the paperwork, to plan and communicate with others using the linear language of words and sentences. Feel the energetic resonance, or emotions, of this activity in your second center lower heart. Make it happen using your first center physical body.

And to the degree you do so, to that degree you will feel amazing. To the degree you are not doing this, to that degree you will feel emotions of dissatisfaction, frustration, anger, anxiety, and depression, even hopelessness and rage. You will feel the bioenergetic feedback of your activity as vibrations that resonate or not, to fuller or lesser extent, with your quantum self.

Initial Actions/Tools for Coping

- Here's another PEMS exercise to try: whatever moves you and holds some deep importance or value for you, write it into the related section (physical, emotional, mental/intellectual, and spiritual). Examples of physical values are money, safety, security, a house, or a car. Emotional values include friendship, marriage, having children, and those things or activities that you desire for yourself.

  Mental or intellectual values would be that which particularly interests you, that which you enjoy talking about—whether certain conversation topics or movie or book genres, and that which you would feel inspired to do as a career. Spiritual values are those activities that help you center yourself, help you find your peace and joy, raise you up into higher vibration, reconnect your senses with God or the Universe, inspire you, and bring you purpose, as well as those tasks for which you feel a higher loving responsibility. When you have finished writing down all of that which holds value or importance for you, take a big-picture view of the themes emerging on your paper.

  Journal about the various themes that emerge from the above exercise, noticing which seem to be more related to a sense of purpose for you. Circle those. Compare them with how you are actually living your life.

- Prepare to listen, then ask within, "Which is my highest purpose? What am I here to do?" Allow answers to bubble up within your higher mind as intuition. Write them down without editing them.

As a variation, in a meditative listening state, ask the Universe to show you a sign, image, or symbol that conveys what you are here to do in this lifetime. Notice what comes in. If you need clarification, you can ask for another sign, symbol, or image to elaborate on what you have been given. At the end of your meditation, journal about what you received, without editing it. Reflect on which of those activities you received in your communication that may be more doable sooner, rather than later, on your path.

- If meditating is too difficult, ask yourself, "If I were CEO of the planet, which of the world's problems would I focus on immediately for change?" Whatever activity arises in your mind, write it down. When you are done, reflect on what you have written. As I've heard Dr. Richard Bartlett say, "Notice what you notice." It is no coincidence that you notice what you notice. If you notice it, it is likely part of your energy stream to notice it and to take responsibility for some activity related to it. When you align with that, you will likely find a surge in the sense of meaning in your life and the purpose for which you incarnated.

Other ways to get in touch with your purpose could be to:

- Give your life meaning and purpose by choosing an activity to take responsibility for, then begin getting involved. If you get stuck on knowing where to start, research who is already doing what in your area of involvement; zoom in on the persons or organizations that resonate with you the most, then see if you can volunteer or work with them. Allow your activities to

evolve over time, in tandem with the unfoldment of
your centers of consciousness.

- Think of career interests as you would recreational
  interests. Follow your heart. Do what feels full of love
  and joy in the present. Anything you do out of love
  is special. When you feel pulled to change directions,
  you can do so . . . gradually, as needed. There is no
  one career that is your purpose. There are many probable vocational paths from which you can choose.
  Change it up as often as you like!

Feeling Stuck Regarding Higher Creativity

You keep having really creative ideas pop into your head,
activity of your higher fifth and sixth centers. It feels as if the
Universe is regularly downloading them into you from the
Cloud.

As you think of these ideas, they send electricity throughout
your entire body. Your fourth center higher heart feels expansive, alive, and energized like nothing in your regular routine.

But your third center self-talk paralyzes you, internally
barking at how inadequate you are for the task. Telling you
how risky or life-changing it would be to head in the creative
direction instead of the usual direction. Telling you that you
will fail. And how much you would disappoint your parents,
who don't understand you and want you to take the safe road.
Or how you might get ridiculed for this cutting-edge line
of work by those who just don't get it and who control the
guidelines used by your professional field.

You may lose your voice often, get a sore throat, have to
cough, or clear your throat regularly. Or your neck tenses up.
Your thyroid functions erratically.

A woman in her forties envisioned making a movie, one that would bring to life an esoteric version of Earth's history and evolution but would be posed as fiction to get people to really think about such possibilities. Her creative ideas were fascinating, and she seemed to have the essential skills for making them a reality. Yet she was stuck from moving forward because she had a hard time believing she was strong enough to go against her mother's wishes for her life, which were for her to become an engineer.

In another case, one young man came to see me, recently having been diagnosed with liver cancer. He wanted to learn healing types of imageries he could do as part of his holistic approach to treatment. During the course of our work together, it came out that he was really of much higher consciousness than his family and especially wanted to travel and work with causes important to third-world nations. He wept as he talked about it, increasingly open and longing for this other way of living. But stuck in his rigid thinking and not even allowing himself to seriously consider a change, he pointed out how he couldn't do any of it because he would be considered a disappointment to his parents, a linearly thinking neuroscientist father and a worrier about his safety and security mother.

Initial Actions/Tools for Coping

When you feel stuck living a life that follows someone else's rules, and you know you are receiving intuitive guidance on some creative and novel direction, remember to be patient with yourself and take baby steps.

- Journal the ideas that are coming to you, the visions that you sense. Write down your ideas and visions without editing them.

- Don't let the big picture overwhelm you, as it tends to for most of us. Break it down into small and doable tasks. Take two or three tasks at a time and budget time in your schedule for them, particularly in the various initial phases of research, fact gathering, getting to know what is already out there, and what sources of funding are available.

- Assess what skills you have and which parts of the creative project would be your part. Don't leave out assessing how much energy and time you have to devote to the task. Be realistic with yourself.

- Consider the team you need to put together to move forward in any given stage of the project. And begin networking or reaching out to meet potential team members. Assess your own skills, and make use of the skills of your trusted team members. Feel like you're an artist but hate the business/marketing aspect? Subcontract that out to someone who loves to sell art and will do it on commission. Don't know how to finance the big project? Seek sound financial advice. Don't know the legal side of things? Ask for help. Remember, perfection is achieved through our diversity.

- Realize that some projects can be started, or even completed, on the side as you continue to work your current job. More involved projects can be transitioned into gradually.

- Remind yourself that when you are about to courageously move forward with some task the Universe seems to encourage you to take, the miraculous and the synchronous seem to come in to play. Ask to receive all needed help and expect the unexpected, no matter the twists and turns.

Attracting the Law of Attraction

You keep hearing about the Law of Attraction, and you're curious.

The Law of Attraction is the universal law governing how "like attracts like" at the level of soul consciousness. Your soul plays a key note that attracts other notes (energy frequencies, thoughts, cells, people, etc.) to its song. Your inner essence that is of loving vibration is sensed by others resonating at a similar frequency.

When you go into fear mode, you slow your frequency and create an energy blockage for yourself. Those who are similar souls to you will not be able to "feel" you as they can when you radiate the love that you are. On the contrary, those who are of similar fear as you will find you. In this way, when you put out negative energy, you will seem to be surrounded by it; when you put out positive energy, you will seem to find it magnetized to you as well.

If you have experimented with the Law of Attraction, you may have noticed synchronicities in your life that seem way beyond chance happenings. But you can't put your finger on what seems to initiate the Law in your life. Perhaps you have even read many books on the topic, but have had inconsistent results.

Here is a client story that illustrates the oddity of the Law of Attraction.

One of my higher consciousness clients came to me for help, wanting to recover from trauma, to leave drugs behind for good, to move into healthier relationships, to set better boundaries, to get the courage to do what her heart moved her to do, and to find her voice again.

She purposely had searched for a therapist who was transpersonal minded, and she enjoyed working with the Consciousness Map because it gave her a way to see the various parts of herself that were in conflict with one another. Her first center had experienced physical and sexual trauma and therefore longed for the sensation of the drug highs, yet also craved natural healing with foods and nature as her medicine.

Her second center was so sensitive to energy that it also liked the drugs for the numbing effect on her negative emotions and overwhelming psychic energies; but numbing to emotion led to the creation of the boundary problems she now had, because she needed to be in touch with her emotions to set healthy boundaries in the first place. And since her trust had been repeatedly violated, she had to work on cultivating trust again in her significant relationships.

The third center part of her needed to get its personal power back and longed to achieve in the field of counseling. Her fourth center heart loved—resonated with—the earth, nature, and crystals. She yearned for healing for the planet and wanted to use nature in her eventual practice of psychotherapy.

Her fifth center felt empty, having lost her voice years prior when she was rejected by her young schoolmates who didn't understand her higher, intuitive knowledge that rocks are conscious, that Mother Earth is a conscious living being, and that the planet is on the verge of shifting into the next dimension (which would be the equivalent of us moving up to the next center as our predominant way of life).

Her Third Eye was naturally wide open, and she received many psychic visions and premonitions. Through her seventh center, she routinely meditated and had had several out-of-body experiences.

We used PEMS and other exercises to help her remember all the aspects of self that she had pushed back into the recesses of her mind.

In our most recent session, we helped her take a conscious step into the Law of Attraction and to get past more of the blocks that had held her back from even asking the Universe for help or setting any sort of intention for her life or career.

She was able to process the unconstructive self-talk that was keeping her from speaking her voice out into the world that needed her so badly. For her, beliefs that she *didn't deserve to ask,* didn't deserve to commune with God or the Conscious Loving Universe were keeping her stuck. *To ask* Source for her dreams to come true, to set intentions for her life, was to put her needs above others, which *was selfish and self-serving.* Underneath those beliefs, we uncovered even deeper seated beliefs—that *there is not enough for everyone* and that *not everyone's dreams can come true,* so it would be *unfair* for her to ask anything for her own life when there is widespread poverty surrounding her all over the globe.

This is pretty typical thinking for someone moving into their fourth center. But we pushed her past these thought themes that were lingering and getting in her way from expressing at her fifth center level.

With help, she was able to see that anything and everything for which she was asking, setting intentions, and dreaming was really for humanity. When she realized that it was a *collaboration* she was asking for, she slowly and mindfully pondered, "Oh, that feels much better in my body; I *feel* more support in what I'm using my voice for." She paused to reflect. "It's a spiritual goal request, a group request, the stuff that changes the world. [Pause] Everyone deserves to have goals met, and that's what I help others do. I encourage my clients to achieve their goals, so I guess I can encourage myself as well. [Pause] When I encourage my dreams, I can show them how to encourage their dreams. [Pause] I think they like my voice because it's different. [Pause] I should use it. Use it!" She encouraged herself with an exclamation and a smile. "The *collaboration* piece uplifts me. It shifts me. It motivates me to use my voice."

For her, the quantum physics language especially helped her past the "there's not enough for everyone" belief. I told her to consider, "What we see as space and nothingness is really a quantum vacuum field of potentiality out of which all material reality emerges as an act of consciousness, awaiting us to ask and set intentions for the group, for the all of us. It is our 'collaborative-loving-group-spiritual request' state of consciousness that works to manifest the something out of the apparent nothing.

"Such group consciousness, making loving prayers for the world, is a potent energetic intention (coherent field of energy) that sets off a chain of energetic responses that help make manifest such requests. In this way, we join in creating what we see around us. We can model this for others. We can be the example that shows others the way to change the world and the vibration of this planet."

She had me write down a lot of notes for her about the

quantum physics piece so that she could go back and read it every time she needed to remember. She was amazed to hear that any "thing" is really more than 99 percent space anyway and seemed intrigued to see the sixth episode of *Cosmos* that I recommended she watch, where a speck of dust was proportionally the particle stuff but the cathedral surrounding it was proportionally the space, in any atom or in any thing that we zoom in on.

I added, "To be in touch with space is a constructive thing. It is your eternal essence. Space is pregnant with potentiality and possibility. Your possibility. Your energetic potential. And according to physicist/chemist W. J. Bray, there is more power in a fifth of a teaspoon of the quantum vacuum of space than in the entire known physical universe. So *nothing* is really *the source of all something.*"

Energized, she decided she was ready to create a vision board for herself. She was prepared to consciously participate in the Law of Attraction by setting her intentions for her more spiritual career direction. Suddenly it dawned on her that she *had had all of her recent goals* listed on a white board in her bedroom and that she had just erased them *because they had all come true.* She paused as she realized the implications of this . . . of her writing them out unknowingly as intentions and seeing that they *all* had recently come true, and in even better ways than she had dared to imagine. This inspired her to take her dreams to the next level and to put them out there on the white board.

I handed her the clipboard so we could help her list her new set of intentions, synthesizing all of that which she wanted to come through her fifth center voice for her career.

After all, at the age of thirty-two, she was starting a master's degree in counseling psychology in a few weeks and was currently working as a volunteer at a camp. On the paper, she put the following phrases: nature psychology, nature spirits, acceptance of others, connection, community, spirituality, teaching, presenting, sharing, increasing frequency, healing trauma, working with those who have autism and have a gifted mind yearning to be released, psychic powers, shifting the planet.

I also noticed that next to *teaching, presenting, sharing* she put a *circle with a dot* in the middle. I smiled to myself because months earlier, we had helped her work through her trauma of speaking in front of a group. When we talked about it then, I had her think of teaching, not in an egotistical way but in a spiritual way. A way in which as she presents, she focuses on the mindset of sharing what she knows and not trying to make it any more than that.

At that time, I drew the circle with the dot in the middle as a symbol of humility and spiritual leadership and as a mandala or symbol on which to focus her energies and mindset. The symbol of the circle with the dot in the middle served to remind her that from the middle we help the group of all of us move forward. We are part of the group that is the circle of connectedness. From the middle, we know the pain of the group because we are there and are in touch with it. From this middle place, we serve, we are humble, and we move forward as a group. The circle with the dot in the middle, this symbol of humility, had already significantly helped her to move past some blocks. Now she was clearly ready to take a few more steps.

On her paper, she also drew a *path.* She was ready *to move forward,* to get her voice back, to stand in her fifth center of consciousness and draw energy and support from it.

Now I have to tell you what happened a few days later, yet another series of synchronicities that never cease to bring tears of joy and validation for this kind of work.

I had just written the sentence above about her drawing a path on her paper when my phone rang. It was Sunday mid-morning. I usually have my work cell off on the weekends, a way of setting energetic boundaries for myself, but I'd had to send a text to my friend a few minutes earlier, so the phone synchronistically was still on. As it turns out, this client had started to text me while I was apparently texting my friend, but then it occurred to her intuitively to "call Valerie," so she thought that she would try to call instead and leave a long message with what she had to say.

She was startled when I picked up. She said she was really feeling the need for me to know what had just transpired, because it was so relevant to what we had talked about on Thursday. She was still shaking, her personal boundaries a bit rattled, and she said she needed help coping with how quickly, and the manner in which, events had ensued.

I had seen her on Thursday afternoon. By the end of Saturday, *less than forty-eight hours after* she had set her new intentions for group good, every phrase on her white board had manifested in her volunteer work with her camp clients, much to her shock and amazement. Her account is summarized as follows:

She had left our session and had gone out to buy *The Three Waves of Volunteers* by Dolores Cannon, a book that I had felt she would resonate with. She was letting it all sink in—our entire talk about Law of Attraction, setting intentions, getting past old beliefs, reframing new thinking about collaboration, her hoped-for career direction—when she went to her volunteer job on Saturday morning. (Keep in mind that she had just intended what her preferred client work would look like. That is, she was inspired to work with clients considered to be developmentally disabled and autistic, with trauma histories, and with whom she could talk about energies and use nature/spiritual psychology.)

That Saturday morning, she had counseling appointments set up with two young men, one developmentally disabled and the other autistic, both of whom where there to process trauma. These were clients she had been working with for three months, and all of the work to that point was pretty conventional trauma counseling. However, things were about to dramatically change in the direction of her intentions.

The first client arrived and saw a Barbara Brennan book on her table, *Light Emerging: The Journey of Personal Healing.* He asked if she were spiritual. She said yes.

Apparently, it was as if a light switch turned on within him. He opened up like never before. He told her that he had never found someone who might understand what he really wanted to talk about. And there he was before her, speaking the very words that validated exactly what she and I had talked about two days prior. He disclosed that he was psychic and very spiritual. "There are spirits everywhere," he told her. "I know," she nodded. He said he could tap into all of that and could connect her to her spirit guides if she wanted. He stepped swiftly, without asking her, into giving her more of a psychic reading, "You have clients with autism . . . they are very spiritual and psychic. They see energy. They see your aura and want to connect to it . . . Your grandfather is trying to talk to you . . ."

"Stop," she quickly interrupted him, and switched to a conversation about energetic boundaries. (I had recommended the book by Cyndi Dale earlier in our work together.)

She told me she had felt uncomfortable having him read her psychically, knowing how much the field of Western psychology stresses that therapists should not reveal personal aspects of themselves to their clients. She wanted to set that professional boundary with him.

But the language of chakras, centers of consciousness, auras, and energies was precisely what he related to, and he requested that she be his teacher along this line of topics (exactly the intention she had set on her white board as to her career and highest purpose, serving exactly this group of clients).

"I think those with autism and Down syndrome like I have . . . have gifts that others can't get . . ." he began again, imploring her to understand what he wanted to reveal to her now that he felt safe enough to do so.

I had chills running throughout my body as I listened to her tell me what he had told her. It confirmed just what she and I had theorized two days prior, that people mistakenly think that those with developmental disorders are not really with it or connected, but they are. Perhaps more through the heart and the right brain antenna than the left.

Twenty minutes after their session ended, her session with her autistic client began.

"Ed is under your chair . . ." he mentioned as he walked into her office. Ed was a spirit he saw.

"What is going on?" she wondered to herself, briefly forgetting about her recent intention-setting moments.

To repeat, these clients' counseling sessions before this morning had involved only very traditional psychology. Yet synchronistically, the other client had started talking about energies, had just announced matter-of-factly that spirits were everywhere, and psychically told her (he did not know her other clients) that she had a client with autism who could see spirits. Now this next client, diagnosed with autism, was indeed pointing out spirits in her office.

This morning was definitely taking a turn in the direction of the Law of Attraction. She was receiving validation after validation of her career aspiration and intentions. Validation of what we had been talking about related to the therapeutic needs of this group of clients, of our theory about their psychic and intensely spiritual nature, which was the result of her previous experiences with other autistic and developmentally disabled clients. Validation about their preference to use the language of energy to talk about their experiences. Validation that the Universe clearly hoped she would pursue this novel line of work.

And, as also synchronistically predicted by the previous client, it did happen that the following autistic client apparently saw and felt very connected to her energy field and had difficulty restraining the spiritual love he felt for her. He uncharacteristically kept trying to hug her, something he had never done before today, and she had to firmly set and help him learn appropriate boundaries.

Was there a difference in her energy field since Thursday that would make him behave so differently from usual? She wondered . . .

Wow! Thursday she cleared her blocks to setting intentions, then set what she thought might be a five-year-plan for her ideal career work with her ideal group of clients. And by Saturday, every word and phrase on her vision board had come into play within her work at the camp.

It did mess with her mind a bit. And we did help her cope with her new quantum reality.

Initial Actions/Tools for Coping

Keep in mind that it is soul energy and quantum consciousness that attracts in this way. Not ego. Ego is working at slower electromagnetic energy levels. Soul is working collectively, synchronistically, nonlocally, holographically through coherent torsion fields that spiral us all together as Love.

- When you find yourself having to work really hard and are clearly not in the Law of Attraction, take note of how ego may be your predominant consciousness. In ego, we are out for our own edification.

  Then, to return to soul energies, regain group awareness. Notice what would be best for the group (whether family, church, organization, community, nation, or planet) in which you find yourself, and align your ego's energies in serving the group good. As you do, you will likely find yourself back in the flow of synchronicity that characterizes the crystalline coherent energy of the Law of Attraction.

- Meditate on the energy of higher love—that is, unselfish, unconditional love—in any way that you can. Some people think of their pets, because it helps them pull the focus of their consciousness up into the energy of unconditional love. Others think of their children or spouse. Some focus on a spiritual figure, like Christ or a saint that for them embodies Divine Love. If the focus of your awareness or imagery helps you lift into the tangible sense of unconditional love as an energy vibration, then it is a focus you can use to lift your ego into the soul mode, which is love consciousness.

- Since it is coherent or well-organized energy that holds and sustains the power to attract, you will have to learn to "hold steady the light" of your soul. If your state of soul consciousness quickly scatters and dissipates, then practice by meditating on holding the energy of love as a focus of your meditation for increasingly longer periods of time. Start with a goal of five minutes. By the time you are living as a soul, you are holding soul energy predominantly throughout your days and nights, even in the midst of chaos. You are living the love that you naturally are.

Chapter Seven

Subtle Energy Imbalances and Emotional Boundaries

Let's now deal with what it is like to feel energy very perceptively. It can overwhelm us at times, especially when we aren't yet sure how to achieve energetic balance. It can also be confusing, for we can have difficulty distinguishing our own energies from others' energies or from the myriad energies emanating from the many things around us.

Energetic/Psychic Sensitivity
Empathic Sensitivity

You tangibly feel the emotions of others around you. When they are angry, you feel the anger come into your body as if it were your own. When they are stressed, you feel it, because your heart begins to race, your breathing quickens its pace, your stomach churns, and your body starts to feel jittery all over. When they are sobbing with depression, you feel your heart get heavy with the overwhelming sadness.

It's confusing. Oftentimes, you're not sure why you feel so anxious, because you can't see any reason for it in your own life. It may not occur to you that it is not stemming from you. When you see suffering on television or in the movies, you suffer. When you see someone kill an animal or another human being, you feel pain.

This empathic or emotional sensitivity takes a toll on your body. You may develop medical issues, such as digestion problems, chronic breathing conditions, or heart pains for which the doctors can find no physical cause.

This subtle energy perception may create problems in your life or relationships. Your family may complain that "you are too sensitive" and that you cry too much. You may instinctually react to the psychological pain you are witnessing, and others accuse you of being dramatic or traumatic.

As you've come to learn, everyone around you emits emotions as waves of energy at a certain frequency or speed of vibration. When your field is overly open or attuned to those persons, you may begin to resonate and entrain to their energy, which is why your heart will suddenly pick up to their pace. When it does, you feel what they feel because you have become as one with their energy field.

When waves of energy resonate, and therefore entrain or exchange patterns of information, they become a collective field of energy. In this state, "you" have become a "we" with them, and their energy is your energy. When this is for intuitive purposes (a fourth center of consciousness perception), there is usually no problem. When this becomes a habit of enmeshment or overidentification with others (second center), problems arise.

One of my Native American clients said that he could not really get into doing the homework on refocusing his thoughts. But he related to the energy imagery and found that it worked quite well for him. Especially to balance his empathic sensitivity, he practiced the qigong meditation of creating the violet egg of protection, in which he would see his auric field as lovingly surrounded by an egg-shaped energy field of violet rimmed in gold. He said he felt calmer and felt that others' energies didn't bother him like they had previously.

When I suggested the violet egg of protection imagery to another client, he got a surprised look on his face. He said that exact imagery had come to him in one of his meditations.

Initial Actions/Tools for Coping

- Practice setting *energetic boundaries*. A good book on the topic is *Energetic Boundaries* by Cyndi Dale.

- Set a new intention with your *mental imagery*. Envision your entire auric field as firm *and* appropriately permeable. Think of seeing a cell that holds its structure and is in healthy communication with other cells. You are this cell in the larger cosmic body. In this state, you are holding steady your light. Your field is clearly organized and coherent, like a tornado or laser of love energy.

  Its outer membrane is still permeable and open to receiving information from others, but your field is no longer overly permeable or so adaptable that it takes the shape of others' energies, and you lose your own felt sense of self. Think of your physical body, how it retains its shape and how the skin is firm yet can exchange oxygen and other substances with the environment that surrounds it.

- *Feel* how far *your field* extends outward from you. Have someone walk slowly toward you and tell them to stop when you feel them in your field. Make a mental note of the distance. For some of you who are very sensitive, you may feel them even fifteen feet away from you.

- For those of you who have a field that tangibly extends beyond the more common distance of three or four feet from your body, practice *"rolling it in."* Kind of like you have an awning that is extending out from the house by fifteen feet, and you roll it in so that it is now tucked in next to the side of the house. I may do this when I have a client that I can tell is sensitive to feeling my intense field of energy; I feel the fringe of my field and visualize myself pulling it in close to my body, even as close as six inches.

Collective Grief/Emotion

While I referred to empathic sensitivity as feeling the emotions of the people you are physically around, I'm using the term *collective emotion* to refer to the emotions of humanity and the planet as a whole that you are perceiving within your own body, nonlocally and regardless of the physical distance.

Though it can be collective anger or anxiety or any other emotion, collective grief is a particularly common struggle for my intuitive clients. I'm seeing it more and more in my practice, and I recognize it when it is described by my empathic and intuitive clients as confusing, out of place, uncharacteristic of or unrelated to their own personal life experience.

Here is one story that illustrates the way it feels and what we can do to shift it.

An unmarried client in her late twenties recently needed a way to break up the heavy and compacted grief energy she felt throughout her entire torso. She had found herself crying profusely, such that it felt to her to be disproportionate to what she expected if it were all her own emotion.

As it turned out, several things were going on. One, she was serving as the primary support for a group of young friends at work who had just lost their spouses in a tragic car accident. She, being an empath, believed she had to hold onto their grief energy to relate and help them. Two, the entire experience had triggered within her a deep loneliness from not having ever felt the love of a partner in the way these women had. She longed to feel loved, in a physical, tangible, visceral way. Three, being of very high consciousness, she could feel the collective grief of the planet. Normally, she could manage it. Now, combined with all of the other sources of grief, it overwhelmed her physically and psychologically.

To help her, I first had her picture the grief as energy and had her use imagery to move it outside her body where it could no longer overwhelm her. Then we had her challenge her belief that she had to hold onto the energy internally to learn from it. She discovered through an experiential exercise that she was able to retain the intuitive ability to access the information from that grief energy without holding it within her body.

Also, she was able to use mindfulness of the body and prayerful imagery to replace the grief and loneliness with an incredible sense of Divine Love. She cried with relief and appreciation for being reminded that the type of connection for which she yearned didn't reside with any physical being but could be accessed through her own focused state of consciousness. For her, a type of transcendent meditation was key to helping her remember that she is never really alone; she told me she actually heard the words, "You are not alone."

Initial Actions/Tools for Coping

When you feel a sudden change in your field that seems to be a thick or overwhelmingly uncomfortable emotion, *reflect on whether it is your emotion, someone else's, or collective emotion* that you are picking up with your empathic "radar." There are many different methods for releasing these emotions, so I recommend trying several to see what fits for you.

- Offer a *prayer* to the community, humanity itself, or the planet at large. Ask for those hurting to be healed and filled with Divine Love.

- Physically move the energy from your body through joyful exercise and play.

- When you notice that you have taken on a distressing collective emotion, *give it a mental picture.* You can see it as a certain color, for example, and you can imagine what it feels like as you set it free outside your energy field.

  When you set it free, you also have the opportunity to cleanse it. In your mind's eye, you can see it as *being purified* with white or golden light, then dissolved back into the quantum field of potentiality.

- See the collective emotion as a *field of energy* with a certain color that is transformed by the power of Divine Love. If you see Divine Love as gold or white, see that color shifting the grey sadness energy, the black hatred, or the muddied hue of envy.

- Imagine Divine Love coming into you through the top of your head as you inhale; as you exhale, feel it radiating outward and into humanity and the planet. As you do, feel your field cohere to the vibration of love energy and sense its power to entrain and transform others.

- Imagine the collective emotion as a *sound* and pierce it with the sound of harmony. Hear the angelic choirs using their voices to transform cacophonous noise and uplift this plane of lower frequency emotion into a symphony of Love.

- However you choose to do it, *strengthen your own field* so that it is coherent. Remember, coherent energy has effects on scattered energy. But if your field is scattered, it is susceptible to being swept up into collective energy that is more coherent than your own.

Ghosts and Foreign Entities

You have always seen "shadow people," ghosts, or spirits and are afraid to tell anyone for fear you will be diagnosed as crazy, though you know you aren't.

Or you have only more recently begun to see discarnate or "foreign" entities and are presently truly worried that you *are* crazy. You feel you need to talk to someone but aren't sure whom you can trust.

Or you kind of like the fact that you are psychic and "see dead people" but wish you could turn it off sometimes.

You used to see "little people" or other-dimensional beings but don't anymore. You wonder whether or not you were imagining or dreaming it but are pretty sure you weren't.

As you grew up, people told you stories of when you were young that sounded like you were possessed because you seemed to have an uncanny physical strength. The thought of that frightens you, especially because there were apparently many witnesses to those episodes. But you don't know what to do about the information.

You have weird mood swings that don't feel like they are your own emotions. Especially anger or rage. This may have started after chemotherapy, after an angry or traumatic time in your life, or after using mind-altering substances.

Your psychic ability to see ghosts or foreign entities is eso-terically described as stemming from your second center of consciousness and is related to empath energy. It is generally explained as the result of deeply tapping into the frequen-cies of the astral/emotional plane, where reside disembodied entities who resonate in consciousness more closely with the emotional-material plane than with higher planes of vibra-tion. When you are resonating with that plane, you pick up information from that plane.

Also, people who are regularly down, depressed, angry, suicidal, or high on drugs tend to attract such foreign ener-gies, as do traumatized people who wish to be out of their body. Disembodied entities seem to be attracted to the inten-sity of this type of negative and very physical energy.

It also appears that if you "beam out" or dissociate from your body, you open the door for other spirits to "beam in" and make use of your energy or body. Unlike in the movies, those foreign energies that step in may be benign or even think they are helping you. However, the sensation of "step-ins" in your field will rarely feel good to you and may cause you symptoms that end up diagnosed as a mental health disorder.

According to Robert W. Alcorn, MD, holistic psychiatrist and shamanic practitioner, he and similarly trained energy healers can help you cleanse your field of unwanted foreign energy through a form of compassionate depossession. I first heard of Alcorn when I attended a video demonstration and workshop by him at an ISSSEEM conference. I watched him work, apparently successfully, with a young woman who had been diagnosed with mental illness and had been institutionalized for a long while.

One of the entities we watched emerge from her field was her grandmother, who thought she was protecting her granddaughter from a traumatic situation. Another was a little boy who had passed, who had seen her in the neighborhood when she was about five and decided she looked safe, so he would hang out in her body.

I have never seen a possession like that depicted in the movies, though I have referred clients with complex situations to a local doctor who also does her version of compassionate depossession. All but one described her as being incredibly helpful, even after just one session.

Really, most of my clients have been able to easily cleanse their fields themselves of unwanted foreign energies with a few simple tools I have suggested.

Scientifically, it is believed that your paranormal abilities involve nonlocal and holographic torsion fields. Your abilities defy the laws of conventional physics and are therefore denied by such traditional physicists. However, MIT- and Princeton-educated physicist Claude Swanson, PhD, has written in his book *The Synchronized Universe: New Science of the Paranormal* about the "new" physics that does explain such experiences. According to Swanson and other similar scientists, paranormal abilities are well documented scientifically.

One of my clients, a teenager, was so depressed that he was failing high school as a result and had resorted to home-schooling. He could no longer be a part of the football team or any of the other social activities that made his life worth living. His father was worried that the boy might attempt suicide if he kept his current path.

After the teen began to trust me, he revealed to me that he saw ghosts. He was overflowing with anxiety that someone would find out, think he was crazy, and commit him. The depression was from carrying that weight for too many years, and he was not holding up well. I went to my bookshelf and pulled down Claude Swanson's books. My reply was to reassure him that even MIT physicists have documented the paranormal. Then I asked him if he had ever tried to talk to the ghosts to assess what they wanted from him. He looked at me in astonishment and answered no. I gave him homework to try it.

After that, his depression and anxiety lifted, because seeing ghosts was no longer an issue to fear. I convinced his father to allow him to simply take the GED so that he could start community college that following semester. He passed the first time without even having to study.

One of my young women was also a bit freaked out because she saw ghosts. She tried the exercises I gave her and was so successful with them that by the next session, she moved on to other topics, because her psychic abilities were no longer a concern. I never cease to be amazed at the healing power of support and validation and a few simple techniques.

Another of my clients believed he had to keep his field wide open to stay in touch with his sister who had recently passed. He couldn't see her, but he said he could feel her and believed he intuited some communication from her. He was clearly suffering symptoms of poor energetic boundaries and agreed to do the imagery exercises I gave him.

As he practiced them in session, he commented how he had never felt so calm and peaceful and grounded. He also agreed to visit the following week the local doctor who does compassionate depossession and was relieved to hear from her that he did not have any other foreign energy within him as he had feared. As such, we were able to rule that out as a possible cause of his recent mood swings. He also agreed to follow up with a local neurofeedback specialist who confirmed my client's psychic brain wave pattern.

Initial Actions/Tools for Coping

If you are having severe symptoms, seek out an *energy healer* known to have success with some form of compassionate depossession.

Otherwise, consider that it is said to be a universal law that foreign entities can only come into your energetic space if they are not forbidden from being there. So *ask them what they need to tell you, request that they leave, close the energetic door to your field (and your bedroom, home, etc.), and send them into the light with the highest of love.* Try the following exercises to help clear your energy of foreign entities.

- Send out the highest love you can and radiate it outward. Firmly communicate that any entity you perceive is to leave your space; send them away with the highest love you can direct.

- Note the power of coherent thought to holographically project and create the very entity you believe you see. Switch to a higher level of consciousness in your thoughts. Live from the plane that does not include perception of dark entity energy.

- Set your energetic boundaries firmly. *See yourself as a coherent field of light* and of the highest form of Divine Love. When you do, you *raise your vibration* to a level that is higher in frequency than the astral "ghosts"; they will no longer be able to see you, just as we cannot normally see that which is lower or higher in frequency than the visible light spectrum of electromagnetic light.

- *Challenge any beliefs* that you have to hold passed loved ones or any other entities in your field or body before you can be in communication with them. Just as with regular people, you can talk to your loved ones who have passed without them residing in your field.

- It is true that you have to be careful whom you talk to about your psychic abilities. *Discern* who is open enough to listen to you and skilled enough to help you the way you may really need to be helped. As needed, choose licensed professionals who are skilled in holistic, spiritual, and transpersonal concerns such as these. Read their websites carefully and remember that no one can guarantee results, because much relies on you and the state of your consciousness. Most often, they really can serve only as guides or teachers to help you learn what you need to do to attend to your own field.

- *You can also stop all use of mind-altering substances,* including marijuana, even if it is legal. If you are naturally psychic, you are more sensitive than most people to all substances. And substances will usually drop you into a negative spiral that often ends in you being hospitalized for psychotic hallucinations and paranoid behaviors and delusions.

Those who work in transpersonal psychology and/or energy medicine have often theorized that substances can tear the etheric web that is the blueprint field of energy between the physical and astral layers of consciousness. The etheric web is said to be in place to protect those who are not ready to see beyond the physical plane of reality. When you tear the web prematurely, you risk a full-blown psychotic break. As you evolve your consciousness naturally, your soul attracts a different molecular structure into your physical body (as cells die and are routinely replaced), and the web is dissolved as appropriate to your new level of consciousness.

Intuition

You wonder if you are intuitive. You aren't sure how to differentiate your intuition from your gut instinct.

You seem to receive creative "downloads" and aren't sure what to do with them. It may leave you increasingly anxious, because you are pressuring yourself to follow up with or act upon what you keep receiving, but you don't feel ready, worthy, or deserving enough.

It makes you nervous to know things about others. You aren't sure what to say to them and what to keep to yourself.

Intuition is most often confused with instinct in the field of psychology.

While *instinct* is the collective consciousness of all of your cells reacting automatically and as a group or system *within* the physical body, *intuition* is collective or group consciousness that is *trans*personal or *beyond and broader than* the realm of your personal consciousness as an individual human being.

Instinct is like the *most basic "operating system" level of consciousness* that comes already installed with your bodily hardware. *Intuition* is like a *more advanced application* that works in concert with the original operating system and the hardware.

Instinct encompasses the consciousness of the *lower* centers, is a personal protective and defensive *reaction to fear* and signs of danger, and is largely felt in the enteric nervous system, in the area of your stomach close to where the second and third chakra vortices are located.

On the other hand, *intuition* is the consciousness of the *higher transpersonal group mind* and is *void of fear.* It comes in or is received as a "knowing" and is more related to the experience of telepathy among humans than to instinct or analytical thought. It seems to involve torsion fields that communicate beyond but in concert with the electromagnetic signals of your instinctual body.

Recently on vacation, my husband asked if I knew what

certain markings and structures were for on the ground of an old fort. I had no idea and am not that savvy about war history, so I caught myself in my left brain trying to deduce the answer. Immediately, I shut that part of my brain off and announced I would try to get the answer intuitively. "Sliding" came to mind. As it turned out, these structures held the base of cannons and allowed for easy sliding from side to side as cannons were aimed in various directions. Intuition gives us knowledge directly.

Since each person experiences everything uniquely through their own fingerprint stream of energy and consciousness, I will simply share more of what intuition feels like to me. I feel, more than see or hear. Some call that clairsentience rather than clairvoyance or clairaudience. I feel a shift in the "force," and I lean toward it to feel and receive the information that drops into my field of awareness. I could also describe it as a soundless sound, because though my ears hear nothing, the words drop into my mind—but not one at a time. They come in simultaneously, all at the same time, as a complete story received and understood.

I am most intuitive when I am tuned in to my clients. We are together in my office, and I am in synch with them. Clearly our energy fields are in resonance and entrainment, and I pick up information from our collective mind and beyond. An idea enters my mind. A picture impression may form as a mental image. It feels as if an unseen guide is over my right shoulder, dropping information into my energy field. It comes in whole, as a sudden knowing. I have come to trust it over the years; I simply pass on to my client what comes to me, as and when appropriate.

My clients routinely respond by asking me if I'm psychic. I tell them I don't think of myself as psychic, but I do work very intuitively. Most of the time what I tell them seems to be their "next breadcrumb" along a spiritual trail of learning, as their next message in a weird series of synchronicities that has appeared recently in their life.

For instance, I tell them something or mention a keyword that "comes to mind," literally, that then inevitably happens to be the topic of a recent article that they stumbled upon, or part of a meaningful or inspirational phrase of something a friend told them the day prior, or the same subject of a book they've been thinking about reading, or the answer to a spiritual or self-growth question they had been contemplating for the past few days but had not mentioned to me.

This happened one day recently, where with one client I felt impelled to use a "surfing the wave" metaphor. She laughed and told me that her friend and she had just exchanged little surfer icons via texts as a way to depict "going with the flow." Later that day, a client asked me before she left what mantra to focus on that week. I replied, "Let it be." A huge smile came over her face, as she rushed to tell me how that particular statement had been her motto for many years.

With the next client, among a few other synchronous references, I found myself suggesting art therapy, and she exclaimed how she had just been looking into and wanting to incorporate art therapy. Then I began using Van Gogh as a teaching point, and it turned out she felt most inspired by Van Gogh and had many of his paintings on her bedroom wall. I commented that her job seemed to suck the soul right out of her, and she showed me her journal entry where she called her job "soul sucking." I even reached for a metaphor to explain something and decided to use the example of a car mechanic. "Say you have a mechanic named Aldan—" I began. "My mechanic's name *is* Aldan!" she exclaimed. We both laughed at the repeated synchronous topics that had come up during the hour, saying, "Seriously, what are the chances?"

Sometimes I just know things about my clients or people in their lives. It feels more solid within me than a guess, but I try to frame it as a guess or a hypothesis when I put it out there for discussion, knowing it is possible I have misunderstood, misinterpreted, or misstated what I have received. In these kinds of cases, after knowing a person maybe only a few minutes, I might choose to tell them intuitively what I think is their issue, what it stems from, what has complicated it, and which is their direction for progress; then I say, "But let me know if you think this resonates or not."

The intuition I receive is more robust than what I would be able to discern from my analytical brain, and I may not have even heard them relay much about their past or history when the intuitive impression begins to form intact. A dear friend of mine once told me, "It is so irritating when you say these things to me. How can you know more about me than I know about myself?" It was a great lesson for me to learn before I began practicing as a therapist. After that feedback, I tried to make a conscious effort to wait for my clients to get to their own realizations, rather than put mine out there for them.

Medical intuitives usually receive their information in somewhat this way, by pointing the focus of their consciousness toward their client's physical health status, then "noticing what they notice." They can offer suggestions about what may be going on with their clients, often because they feel it within themselves.

One young acupuncturist told me how she went into her next client session practicing the full-body "listening" tools I had demonstrated for her (to get her out of her overanalyzing left brain). She explained how in two different cases that morning she had picked up symptoms or diseases that her clients had not revealed to her, but that became evident as they removed their clothing or were later confirmed in some fashion for her.

For instance, she suddenly felt nauseated when a client entered the room, and it turned out that the client was feeling nauseated and ended up vomiting before the appointment was completed, though the distress had not been mentioned during the initial assessment or consultation phase.

Sometimes these "knowings" or intuitions come in while we are sleeping, but we remember them upon awakening. One morning I awoke with "full" knowledge of how the energies of the seven chakras line up with the seven centers, the seven rays of esoteric psychology, the astrological or planetary forces and triangles of three, the cosmic realms of energy, and so on. It was so clear that I chose not to take the time in the middle of the night to write down what came in because I firmly believed that I could not forget it, any more than I could forget my own name, and because I knew it would take me hours to write it down anyway. (I now wish I had written it down.)

I have come to realize that I may wake in the middle of the night and be momentarily retaining more complete soul consciousness, which contains the intuitive knowing, but that my personality consciousness may not be capable of holding or understanding it all, so it feels as if it leaves me.

I live intuitively. That doesn't mean I am conscious of the intuitive level of my energy field 24/7, because I point the focus of my consciousness on the bathroom when I am cleaning it or on what I may be cooking at the time. However, it does mean that the intuition from my higher centers is what I use to guide my life. Anytime I am making major, and often numerous minor, decisions in my life, I focus in on that part of my energy field. I feel the direction I am pulled toward. I notice the impressions that arise within me. I aim for that.

Initial Actions/Tools for Coping

Each person has one or two physical senses that are more sensitive than the others. The same is true with intuition and general perception of subtle energy. If your sense of sight is the strongest, then your intuition may be more toward clairvoyance. If your sense of hearing is the sharpest, then you may receive your intuition more as clairaudience.

Think of intuition as simply one of your many levels of intelligence and keep it within *balance*. If your "radar" is overly focused on transpersonal realms, you can fall prey to some of the problems mentioned above, such as picking up too much empathic energy or becoming too disconnected from your body, friends, and family, or your daily tasks.

Note that intuition usually seems to be the knowledge that comes in as you are doing your soul purpose. It is therefore very different from person to person, as if we specialize in certain types of intuition based on our highest soul purpose.

For example, if your soul purpose is the psychological healing of others, your intuition will come in more readily in that form. If it is for physical healing of others, then your intuitive radar will be more tuned into that. If your soul's purpose for incarnating in this time and space is to advance science and technology, mathematics, or the arts, then your intuition will be honed to that and you will be called "genius." If it is to communicate with souls who have passed on or who are residing in other realms, then your intuition will be about that. It isn't as common that our intuition picks up all things in all places. It is an individual talent, a gift of soul consciousness, to receive directly and intuitively whatever information we need to fulfill our purpose.

- Practice *discerning the intuition from instinct.* If there is fear involved, it may be your instinct and your cells trying to move you in a way they believe is best to keep your material body safe. But it is not likely to be your intuition.

- Practice *receiving and acting upon* your intuition. There are classes available to teach you how. Everyone has the potential to tap into these subtle fields of energy. It is a matter of practicing to turn the focus of your consciousness in the direction of subtle perception. *Where* you focus it determines that which you receive and perceive; and the more you look, the more you'll see. *How often* you practice tuning into your intuition determines how easily you are consciously interconnected with it on a regular basis. And the more you act upon it with trust, the more you receive. Intuition develops like a positive and upward spiral.

- Practice listening. When you truly listen, your mind is still and open to receiving.

- Consider that *as we act, we receive* our intuition or spiritual gift. The Universe is quite efficient and seems to withhold energy until it is needed. So if intuition is coming to you, then the Universe has already decided that you are one of those who can best do something with the information. Think of your intuitive perception as opportunity that awaits you. Do what you can, as you can. And remember, the responsibility lies with the group of us, not with you alone. Take baby steps.

Being Overwhelmed as a Healing Practitioner

As a healing or energy practitioner, you may frequently feel overwhelmed by the sheer volume of energy you are around from other people, your clients, your family, and collectively.

You are in resonance with your clients and are sensitive enough to pick up their energy, whether physical, emotional, mental, or spiritual. All centers of consciousness are involved, and you feel unsure of how to turn down the volume on your empathic, intuitive, and other psychic abilities.

Often it feels as if your field is contaminated or polluted, thick or muddy, weighed down and not like you. You're tired, even drained, most of the time. You have trouble sleeping.

You may believe that you need to take clients' energies into your own body in order to read or cleanse them.

I used to have a belief that my energy field had to merge with my client's field in order for me to gather the information in their field intuitively. It was just a belief. But what I noticed was that my clients who were sensitive to energy could feel my energy. I could feel them feeling my energy empathically, and I could pick up the signals that my energy was too intense for them, even if I was not saying a word.

So I began to experiment, and I have come to now see (thanks to the science of torsion fields) that I can hold my field close to me and still pick up intuitive signals from their field. And my clients no longer have to put up with my energy comingling with theirs.

Initial Actions/Tools for Coping

Remember the esoteric teaching that *"energy follows thought."* Or, as I've often said, *as* we focus, so we are *there;* that is, where we point or focus our mind, our experience in consciousness follows. And remember, too, that according to fMRI scans, the brain responds similarly to what we imagine we are experiencing and what we are physically experiencing. (If you want, go ahead right now and imagine biting into a sour lemon. Notice how quickly your body responds and you salivate.) The brain tells the body to respond, nonetheless, in about the same way. This seems to be partly why using imagery with our clients seems to work so well and why using imagery to set energetic boundaries works very well for most of us.

- Before your sessions, cohere your own field. *Center yourself* by grounding into your own field with deep and mindful awareness. Some practitioners feel their core point at the level of the belly button. Others feel it in their heart center. Notice what feels more centering to you.

- Allow *white light to come in and cleanse* your field with the highest crystalline coherent energy you can imagine or intuit, both before and after sessions.

- Use *music to reset* your field. Some people buy CDs designed to tone the centers. For me, there is something about the chanting of monks and great choirs singing in the grand cathedrals that feels incredibly cleansing and uplifting.

- *Warm baths* also tend to cleanse and are good for sleep as well. Many of us enjoy the effects of Dead Sea salts or Epsom salts in our bath. Use only natural salts like these, as synthetic varieties labeled as bath salts may contain harmful substances with potentially severe medical effects.

- Do a *loving constellation of emotion*. Take out a piece of paper. In the middle, write down the word love, then circle it. From the circle, draw a ray for each of the following: sights/visuals/imageries, colors, scents, sounds, tastes, physical sensations, environments, thoughts, and any secondary emotions that arise. At each ray, write down keywords that describe that which you associate with love or that which helps you feel love energy.

  For example, you may write down the color of gold, the sights and sounds of the ocean or seagulls, the scent of eucalyptus, the taste of salty air upon your lips, the sensation of your heart expanding, being outside in nature, the thought of feeling free, and the secondary emotion of joy. Note the many sights, visuals, colors, environments, sounds, and scents associated with the highest love energy you can feel. Practice those multisensory imageries as your meditative focus regularly but especially at the start and end of your day.

- Get moving outdoors in nature, which feels incredibly cleansing and balancing. Set your bare feet in the rich soil or grass, and feel your field reset to that of the planet.

Distinguishing Dream States

You may feel overwhelmed or confused by the rush of "dreams" you are experiencing. You may wonder how to interpret them or if you should bother. Some dreams feel quite different from others, but you don't know what to make of them.

There are many types of "dreams."

I think of dreams as experiences in consciousness that may or may not be "personal." Some are an amalgam of subconscious and conscious personal thought, registered and reflected in our mind through the imagery language of the universe. Other dreams are a flight of our soul consciousness to any number of multidimensional realms, captured by the brain radar, seen and monitored as well through the language of mental imagery.

Some of you have simultaneous dreams, in which you feel you are registering more than one dream at a time.

The typical dreams I call "ego" dreams, since the term is generally understood to refer to the personality consciousness. Egoic dreams come from the lower three spheres of personal consciousness. In ego dreams, it is as if you are problem solving or working out your worries and anxieties. The worst of these are your nightmares; if the focus of your consciousness is on seeing the world as scary, that is what you will experience, similar to the minority who experience a scary NDE.

But not all ego dreams are scary. In those that aren't, you may just seem to be working or in some way engaging with people from your regular life. During sleep, you may be in a state that allows for subconscious information that normally is suppressed or repressed to bubble up to the surface of your field of personal consciousness.

"Soul" dreams feel quite different and spiritual and can arise from the transpersonal centers, holographic quantum self, and collective consciousness. In "soul" dreams, you may be traveling to some other dimensional reality, learning something, having something spiritual be revealed. The variety of soul dreams is too broad to cover here, but simultaneous dreams particularly suggest that you are synched into holographic realities and simultaneous "lives."

Just this morning, before writing this section, I awoke from such a dream. In it, I received word of the pope's permission for me to view some esoteric books from the Vatican collection, brought along with him on his US tour. I was led to a small room, where I approached a modest hand-hewn wooden table stacked with numerous books of regal proportion. As my eyes drank them in, my body was electrified with delight.

Two muted navy blue volumes were perched at the very top. One was nearly the length of my arm and depicted on its cover silver-gilded vertical rays of force stemming from a single point of light. Overlaying the book on rays, engraved magnificently in gold with ornate script the size of my hand, was the title, *The Book of Spheres*. Two feet wide and a yard long, the cover only slightly worn for its age, the book's beauty overwhelmed my senses as I beheld the series of embedded orbs, etched in gold and gleaming with a kaleidoscope of colors.

In the section on the intuition, I mentioned another soul dream in which I was being shown the answer to a contemplation I'd had along the lines of esoteric astrology and energy lines of force. In another, I remember being in a crystal palace, where I was handed a box containing an esoteric gift. In yet another, I was handed a Native American type of talisman. I had to look up the word when I awoke, because I was not sure precisely at that time what the word even meant; yet within the dream, I had been told that I was being giving a "talisman."

Such dreams are usually as vivid in physical sensation as this life feels. Some can feel irrational compared to this plane but full of symbolism that you intuitively comprehend at a higher level. The symbolism is personal to us; the personal mind has to convert quantum reality to something more comprehensible by the human brain, and it apparently does so through mental imagery that carries meaning for us.

Soul visits to other planes of reality beyond our own may include visits to other planets or universes or to other planes that are nearly impossible to put into words. It may be that these are planes on which you reside at some multidimensional aspect of your being that is not of this physical plane. I believe that is what is happening in many near-death and "past-life" regression accounts that are not of this world.

Some of you have memories arise within your waking consciousness, or under past-life regressions, and wonder if they are dreams or a past life or similar memories of your higher self.

A woman in her fifties asked me about some dreams she had been having recently. Within them, she knew she was dreaming. They seemed to cover a variety of scenes that her conscious mind could not explain, from an assortment of lifetimes representing different eras of time.

In these dreams, she seemed to be reliving experiences she could only describe as spiritual. She was perhaps hesitant to admit the content of the experiences, yet gave some examples in which it seemed she had abilities she could not explain to the satisfaction of her left brain. She did know that these were not usual ego dreams and felt she was being reminded of something she once knew but had forgotten. She asked me if it were possible to be existing simultaneously in many places at once.

I told her that quantum physics might help us explain what was happening. I described energy in terms of pulses, moving outward as waves of vibration, concentric rings of motion. I told her that we could see consciousness as sets of such energy carrying information, with the largest set being the ocean of Consciousness as a whole, comprising all others.

I explained how energy gets ratcheted from higher and faster moving waves, or frequencies, down to slower and lower frequencies, from torsional fields of energy down to magnetic and electrical currents of the body. She could think of herself as existing simultaneously at these many bands of frequency, just as the colors of the rainbow exist together simultaneously as white light. At every range of "color," or dimension, exists a certain "reality" that is appropriate and consistent with the energetic characteristics or qualities of that particular dimension.

We speculated that time could be seen from the bird's-eye view of Source as a specific localization ("collapse of the wave function," to use quantum physics verbiage) in space or a cohering us into a place we call space, a limited 3-D perception.

Source could "condense" us out of the quantum vacuum field of potentiality into many different places in "time" simultaneously, just as drops of dew condense on the morning grass on several blades at the same time.

I said that we could access (through dreams or meditations) each of these aspects of Self across time and space, because they are all Self, simply spanning many frequencies and dimensions in a spectrum of existence.

She told me that the conversation sent chills throughout her body, waves of recognition of some higher truth.

We went on to help her make sense of her recent history, as a woman in this day and age. We noticed that she'd had higher awareness as early as four years old (she said she could remember "knowing" that her parents were not her real parents . . . and, no, she wasn't adopted). She had blocked it from memory the more she dissociated from her body due to childhood trauma. She had apparently dissociated so much over the years that she became sicker and sicker and had turned to holistic healing because Western medicine seemed stumped as to what was causing her body to shut down and become so weak.

We also helped her see that the sicker she became, the more she turned inward; her left brain was not working very well, but her right brain seemed to serve as the antenna to her higher self. She began having all of the "spiritual" dreams when her left brain was turned off, so to speak. The more of this spiritual self she accessed, the healthier she became. But then she stopped meditating, and she began to feel the sickness coming on again.

Toward the end of our talk, I said "Let me throw this out, like a theory, and let me know if it feels as if it resonates in any way."

I said, "What if you chose to be born in this genetic lineage, to have these specific symptoms when you were ready to wake up again and remember the part of you that exists beyond time and space, knowing that you might become so entrenched in this linear dimension that you might need these symptoms to pay attention? And what if you chose to be born into this early childhood trauma because it could give you the sense of experience you needed to fulfill your purpose at this time?

"Here you are, working all of these years in the very Western mental health field, talking to many people who have come to know you as extremely left brain and rational . . . and now you mention the extraordinary events, memories, and experiences that are surfacing as a result of your illness . . . and it moves these other people to come out to you and to reveal their own extraordinary senses and experiences that they otherwise would never have felt brave enough to talk about in the environment of the hospitals and residential treatment centers in which you work.

"And maybe you are one of the waves of volunteers, those of higher consciousness, who have come to this planet at this time to assist others in opening up and waking up, so that we as a planet might make a quantum leap in consciousness, a shift into a faster dimension of reality.

"Maybe the purpose in your forgetting was so that you could really fit in among those healthcare providers you would eventually come to help awaken, but you needed your illness to make you sit still and meditate on other possible dimensions of reality, to be broken open in your desperation to heal.

"You are a bridge. You are here to help others open up and share these unusual experiences. You are here to serve in the awakening to a higher dimension of reality."

"Yes, yes!" she agreed as she hugged me tightly. "All I can say is *that* really resonates for me. It seems to make it all fit together. Now it all makes so much sense! I think my left brain was in the way, though, of the way my shamanic healer was expressing it to me . . . and I needed this quantum physics language to make some sort of sense out of it all!"

Initial Actions/Tools for Coping

During the state of sleep, your body and its cells are repairing and renewing, but your soul has no need of a "shutdown," so it is busy visiting other planes of reality. According to Bray's writings, we can speculate that these are real experiences of your soul or a higher dimensional aspect of you. Just as real as this life. Consciousness creates all experience by attracting or bending the atomic world to itself.

In *Discipleship in the New Age II* by Alice A. Bailey, we are told: "Everything is a state of consciousness and the passage of time is but the succession of varying states of consciousness." Consider that food for thought when it comes to your dreams.

- *Journal* your dreams. At the very least, journal your soul dreams so that you can remember them and learn how they progress over time.

- For dreams that involve other planets or planes of reality, write down as much *detail* as you can remember. Note any *patterns* that emerge over the years. The same with any "past-life" memories or near-death experiences that arise. You will likely find that others speak or write about the same planes of reality or memories that you experienced, and you will want to be able to take out your notebook to objectively compare.

- When you seek to interpret a dream, there are many ways you can go about it. As a start, you can take note of the *objects, people, and places* within the dream. Ask yourself what *"quality" you associate* with each person, place, or object. Write them all down. Then "read" the dream, *quality by quality*, using the qualities as a symbolic code to be read and deciphered.

It does seem that the Universe speaks to us in symbolic imagery because our rational consciousness can understand it better as a picture. As they say, "a picture speaks a thousand words." Furthermore, in my opinion, it is more useful for you to think of what the symbol means for you (e.g., red wagon = the quality of childhood, being underwater = the quality of feeling overwhelmed) rather than overly rely on what others say it means in our collective consciousness. And again, since ego dreams often mirror the fears we are working through, it is constructive for you to pay attention to your dreams and to allow yourself to see what you may routinely suppress. Healing is always a process of returning to wholeness consciousness.

Balancing Your Energetic/Emotional Boundaries

For those of you who do need to set firmer energetic boundaries, you probably are frustrated that you have a difficult time saying *no*.

Or you believe you say *no* but feel that people don't hear or respect the line you have drawn.

You may frequently feel like a victim in your relationships. You often feel used. Most certainly you feel exhausted, maybe even depressed or generally anxious and resentful. It is like other people suck the life out of you, and they too often literally do.

You tend to make yourself responsible for managing other people's emotions for them and, as a result, know yourself to be a people pleaser, helper, entertainer, caretaker, mediator, or peacemaker.

You feel like the one who does all of the work, at home, in relationships, and at the office. You feel it is easier to do it yourself than to delegate or share responsibility with others. For some, you don't want to say *no,* because it makes you feel more worthy, productive, successful, and so forth to say *yes.*

Or you keep others at a distance. You erect a thick emotional wall around your heart that protects you from being hurt by others; but that wall seals you into isolation, as well. It feels depressing to think you'll never find a partner who understands you and that you will forever feel this lonely. Yet you can't see how you push others away.

You recognize that you pendulum swing from too thin a boundary to too thick a boundary, reacting and overcompensating, inconsistent in the boundaries you set with others.

You can feel the vibrations of all objects, but may not realize that as the underlying reason for being cranky and obsessed about keeping the house clean when it is disorganized or a cluttered mess.

Though you may not have realized it until now, you are sensitive to technological devices and the electromagnetic energies (especially blue spectrum light) they give off—phones, computers, televisions, alarm clocks, other personal electronic handheld devices, and let's not forget power lines, transformers, power stations, and power plants.

You feel mood swings each time there is a significant change in the weather or the seasons. You feel depressed each winter, anxious with shift work, and find it difficult to adjust when you cross time zones. You are sensitive to Earth's planetary cycles and shifts of one type or another.

Restorative sleep is a particular problem for you.

A young mother came into my office complaining of emotional sensitivity and chronic anger. She cried as she told me that she could not stand being around her children because of the noise, even if they were laughing and just being kids. Coping was a struggle, and most often she retreated to her bedroom to lie down and be alone. Thankfully, her cousin was available to watch the children, but it made her sad to not be able to bear being around them. If she pushed herself to stay in their presence, she would find herself easily upset and yelling at them, which she hated in herself. She was tired of Western psychology medicating her symptoms but not helping her to heal.

Neurofeedback was used to assess her brain wave pattern

and rule out traumatic brain injury or concussion, based on the complete symptom picture she painted for me.

We assessed her dissociation level and talked about how often she found herself out of body and looking at herself from near the ceiling. That turned out to be a regular habit, stemming from trauma in childhood. She could do it at will. I suggested she stop and helped her cope in other ways.

Sometimes she had a problem distinguishing what had actually happened in her reality from dream or alternate reality. She could feel spirits and frequently experienced déjà vu and precognitions.

As an initial grounding homework, and to build up her felt sense of energetic boundary, we had her practice the violet egg of protection; it was an auric shell she could use to keep her boundary firm and keep foreign entities from invading. She agreed to notify all spirits that her field was her field, to send outside her field with love any entities currently residing within her, and to let them know that they were no longer allowed to stay inside her.

Next, we used imagery and mindfulness to get her completely back in her own body. She started with feet mindfulness and patiently worked her way up to her head. The qigong exercise "microcosmic orbit" was helpful for grounding her field in her awareness, rotating slowly down from the top of her head, then back up her spine to her head. When she had completed this mindfulness exercise, she reported feeling very solid, relaxed, and peaceful, her emotions down to a 1 or 2 from the 9 she had initially scaled as her distress. And that was just her first attempt at the experiential exercises in our work together.

Another woman in her twenties came to realize that she routinely internalized and overidentified with the emotions from others that she felt as an empath and as collective energy. She could no longer differentiate who she was, and how she felt at her core, from all of the external energies that she absorbed. For example, if she saw a scary movie, and felt (like a true empath) the feelings of the dark character in the movie, she came to believe she was the evil and violence she now internalized and sensed inside her own body.

We used art therapy to eventually help her distinguish her own core from the energy she felt outside herself in others and collectively. She was better able to differentiate, through the use of color, the energies that were hers, feeling them more as varying shades of blues, greens, purples, and oranges. She was also able to see the external energies as different colors in her body. And by seeing it all in pictures, she seemed to better understand which were external yet internalized and which were her own core energies and qualities.

We also had her do an experiential exercise. We had her imagine one of the dark characters with whom she overidentified. To go fully into that which she feared, feeling his feelings and knowing his thoughts. Rather than pulling away this time, judging that she must be this since she feels this, she was able to sit with and follow the imagery wherever it led her. Soon she discovered she was immersed in great compassion and unconditional love and came to realize that it was a state of consciousness not felt, much less understood, by many others. She was able to sit with it more and see in her imagery what happened next. She watched as the great force of unconditional love made the man so full of evil shrink and shrink and then dissolve in the state of love. She felt completely cleansed, an energetic boundary issue healed.

Initial Actions/Tools for Coping

- Take out a piece of paper and *draw* what you believe your energy boundary looks like. One of my male clients drew his boundary as a suit of armor. It was thick, made of steel, and worn tightly against his body. One gal drew hers by not drawing anything other than a thick red wall she erected twenty feet out. Take a thoughtful look at what you draw. Notice if it feels too thick or thin, too full of holes or rough patches, too close or too far away.

  Then draw what you believe a healthy boundary would look like. Imagine it often for yourself. Notice any ways you begin to feel or act differently. Then use *imagery* to shift your boundary to a healthy one. Energy follows thought. Notice how your behaviors begin to shift as a result.

  Discern assertive from aggressive boundary attempts. Practice assertiveness, being your authentic healthy self around others, learning to hold steady your spiritual light. But try not to pendulum swing from overly passive boundaries to overly harsh or aggressive boundary setting. To make it easier on yourself, try telling people what you *can* do for them, in balance, rather than emphasizing what you *can't* do for them.

- If you have a hard time *saying no* to others, then first practice it energetically. That is, in your mind's eye call those others forward. Imagine that their higher selves are gathered with you in a higher plane and that each of you is taking turns speaking with one another. Tell them what you need of them. Ask them what they need of you. Agree to tear up soul agreements that are outdated. Make new soul agreements that arrange for you as a group to collaborate on the larger collective good and to support one another in that effort, rather than to overly focus on personal life lessons.

It is weird how this can start to ignite a change in others in "real" life; it seems to involve those "spooky action at a distance" torsion fields, and it may be that their brain really does pick up what is being communicated nonlocally, as modern experiments suggest.

- Think of your chakras as intake and outlet valves, and allow them to work on your behalf. Imagine *receiving then sharing at every level* of the Consciousness Map.

- Take responsibility for *your own part* in the whole. Allow others to take responsibility for their part in the whole. Adjust your self-talk as needed. Remember that if you drive the lives of others, they don't have to drive themselves. You have rescued rather than helped or facilitated. Get back in your own lane, and drive your own life.

- Notice how the relationships you attract, personally and at work, are *energetic matches*. If you routinely give, you are a perfect match for takers. If you are a talker, you will attract good listeners. If you are a doer, you certainly attract those who need things done. Decide what you want your relationships to feel like. (Remember, in need-based personal or material consciousness, it seems that opposites attract; it is at the level of soul consciousness that like seems to attract like in the spirit of love.)

Chapter Eight

Conclusion

Rest assured that if you resonate with this book, then you have tapped into the intuitions and understandings, the extraordinary senses and sensitivities, the callings and the sense of urgency that is indicative of higher consciousness. You resonate more with love, inclusivity, responsibility, and collaborative cocreation than you do with fear, one-upmanship, judgment, and retribution. It may still be a battle for you to "hold your light steady," but know that you will eventually tilt the scale of balance in love's favor.

It is time we stand out and stand firm as we rise to meet the global challenges of our day. It is time we test our faith in the patience and power of the highest vibrations of love, so that we may dissolve the pervasive anger and mindset of harmfulness around us.

Rather than a long conclusion to this book, I think I will simply leave you with this answer that arose within my consciousness many years ago as I contemplated how to "dissolve" my sense of ego self and merge it with the soul self.

It is about what it is like to leave ego consciousness behind *as the director of daily life* and to give it its higher function as a vessel for soul. It is allegory for the soul as a light pulsing on and off across time and space, a firefly in the Cosmic Night. And for the soul's cloaking into the apparent darkness of the quantum field of potentiality, the great Cosmic Night, and redressing with a new form, one better suited to the higher light it is becoming. Cyclically, infinitely, Life without end. Always in the state of becoming, we are.

How do I dissolve?
It is in letting go of the known,
to adventure into the unknown,
to be
beingness itself.
In choosing nothingness
there is All-ness.
It is the opposite of what the little mind believes,
thinking that in retaining,
power is.
But to "retain"
is to put a boundary on
what can be experienced.
And in imagining that boundary,
more gets left outside of self
than possibly could be held within.
A prison self becomes
until the bars,
shutting in heart and mind,

are lifted,
revealing lightness in the darkness.
Cosmic Night,
where trillions
of fireflies illuminate
its sky.
Piercing clarity.
Flashing on and off
in the cycles of time.
I am but a firefly
in the Cosmic Night.
And yet I AM
the Cosmic Night,
where there is at once
Lightness in the Darkness
and Darkness in the Light.

May you continue to bring your light into the darkness of this slower vibrational dense and material plane, so that together we transform this reality with the united power of our unconditional love.

References

Alexander, Eben. *Proof of Heaven: A Neurosurgeon's Journey into the Afterlife*. New York: Simon & Schuster, 2012.

Atwater, P. M. H. *Near-Death Experiences: The Rest of the Story; What They Teach Us about Living, Dying, and Our True Purpose*. New York: MJF Books, 2011.

Avila, Teresa of, and Dennis Billy. *Interior Castle: The Classic Text with a Spiritual Commentary*. Indiana: Ave Maria Press, 2007.

Bailey, Alice A. *A Treatise on Cosmic Fire*. New York: Lucis Publishing Company, 1982.

———. *Discipleship in the New Age II*. New York: Lucis Publishing Company, 1986.

———. *From Intellect to Intuition*. New York: Lucis Publishing Company, 1987.

Bartlett, Richard, and William A. Tiller. *Matrix Energetics: The Science and Art of Transformation:*. New York: Atria Books, 2007.

Becker, Robert O., and Gary Selden. *The Body Electric: Electromagnetism and the Foundation of Life*. New York: Quill, 1985.

Bray, William J. *Quantum Physics, Near Death Experiences, Eternal Consciousness, Religion, and the Human Soul*. CreateSpace Independent Publishing Platform, 2012.

Bruyere, Rosalyn L. *Wheels of Light: Chakras, Auras, and the Healing Energy of the Body.* Edited by Jeanne Farrens. New York: Fireside, 1994.

Cannon, Dolores. *The Three Waves of Volunteers and the New Earth.* Arkansas: Ozark Mountain Publishing, 2011.

———. *Custodians: Beyond Abduction.* Arkansas: Ozark Mountain Publishing, 2013.

Chapman, Gary. *The 5 Love Languages: The Secret to Love that Lasts.* Illinois: Northfield Publishing, 2009.

Chi Wellness. "Beginning the Qigong Lifestyle." Accessed May 1, 2015. http://www.chiwellness.net/BQL.html.

Chopra, Deepak. *Quantum Healing: Exploring the Frontiers of Mind/Body Medicine.* New York: A Bantam New Age Book, 1990.

Church, Dawson. *The Genie in Your Genes: Epigenetics Medicine and the New Biology of Intention.* California: Elite Books, 2007.

Dale, Cyndi. *Energetic Boundaries: How to Stay Protected and Connected in Work, Love, and Life.* Colorado: Sounds True, 2011.

Dubrovnik Peace Project. "Dubrovnik Peace Project: Scientific Research." Accessed May 1, 2015. http://www.dubrovnik-peace-project.org/sci/maharishi_effect.htm.

Eden, Donna, and David Feinstein. *Energy Medicine.* New York: Tarcher/Putnam, 1998.

Einstein, Albert, and Robert W. Lawson (trans.). *Relativity: The Special and the General Theory.* New York: Wings Books, 1961.

Emoto, Masaru. *The Healing Power of Water.* Singapore: Hay House, 2004.

Gerstein, Lawrence H., and Matt Bennett. "Quantum Physics and Mental Health Counseling: The Time is . . . !" *Journal of Mental Health Counseling* 21, no. 3 (1999): 255–269.

Global Country of World Peace. "Global Peace Initiative." Accessed May 1, 2015. http://www.globalcountry.org/wp/current-projects-2/maharishi-vedic-pandits.

Goldman, Jonathan. *Shifting Frequencies*. Arizona: Light Technology Publishing, 1998.

Goswami, Amit. *Physics of the Soul: The Quantum Book of Living, Dying, Reincarnation, and Immortality*. Virginia: Hampton Roads Publishing Company, 2001.

―――. *The Quantum Doctor: A Physicists Guide to Health and Healing*. Virginia: Hampton Roads Publishing Company, 2004.

Grof, Stanislav. *When the Impossible Happens: Adventures in Non-Ordinary Realities*. Colorado: Sounds True, 2006.

―――. "Revision and Re-Enchantment of Psychology: Legacy of Half a Century of Consciousness Research." *Journal of Transpersonal Psychology* 44, no. 2 (2012): 137–163.

Haisch, Bernard. *The God Theory: Universes, Zero-Point Fields, and What's Behind It All*. California: Red Wheel/Weiser, 2009.

Hanson, Rick, and Richard Mendius. *Buddha's Brain: The Practical Neuroscience of Happiness, Love, and Wisdom*. California: New Harbinger Publications, 2009.

Hunt, Valerie V. *Infinite Mind: Science of the Human Vibrations of Consciousness*. California: Malibu Publishing Company, 1996.

Jacobs, Alan. *The Bhagavad Gita*. United Kingdom: O Books, 2003.

Kaku, Michio. *Parallel Worlds: A Journey Through Creation, Higher Dimensions, and the Future of the Cosmos*. New York: Anchor Books, 2006.

Kessler, David. *Visions, Trips, and Crowded Rooms: Who and What You See Before You Die*. California: Hay House, 2010.

Kuhn, Merrily. "Understanding the Gut Brain: Stress, Appetite, Digestion, and Mood." Seminar sponsored by Institute of Brain Potential, Denver, Colorado, 2014.

Large, Matthew, Swapnil Sharma, Michael T. Compton, Tim Slade, and Olav Nielssen. "Cannabis Use and Earlier Onset of Psychosis." *Arch Gen Psychiatry* (2011): doi:10.0110/archgen-psychiatry.2011.5.

Levine, Peter A., and Ann Frederick. *Waking the Tiger: Healing Trauma.* California: North Atlantic Books, 1997.

Liester, Mitchell B. "Near-Death Experiences and Ayahuasca-Induced Experiences: Two Unique Pathways to a Phenomenologically Similar State of Consciousness." *Journal of Transpersonal Psychology* 45, no. 1 (2013): 24–48.

Lipton, Bruce H. *The Biology of Belief: Unleashing the Power of Consciousness, Matter, and Miracles.* California: Elite Books, 2005.

Maharishi Health Education Center. "World Peace Assembly: 8000 Yogic Flyers Create Coherence in World Consciousness." Accessed May 1, 2015. http://www.maharishitm.org/en/wpaen.htm.

Maharishi University of Management. "Summary of 13 Published Studies." Accessed May 1, 2015. https://www.mum.edu/about-mum/consciousness-based-education/tm-research/maharishi-effect/Summary-of-13-Published-Studies.

Manly, Steven. *Visions of the Multiverse.* New Jersey: Career Press, 2011.

McCraty, Rollin, Mike Atkinson, and Dana Tomasino. *Science of the Heart: Exploring the Role of the Heart in Human Performance.* California: HeartMath Research Center, Institute of HeartMath, publication no. 01-001, 2001. Accessed March 29, 2015. http://www.heartmath.org/free-services/downloads/science-of-the-heart.html.

McMoneagle, Joseph. *Remote Viewing Secrets: A Handbook*. Virginia: Hampton Roads Publishing Company, 2000.

McTaggart, Lynne. *The Field: The Quest for the Secret Force of the Universe*. New York: Harper Perennial, 2003.

———. *The Intention Experiment: Using Your Thoughts to Change Your Life and the World*. New York: Free Press, 2007.

Mitchell, Edgar, and Dwight Williams. *The Way of the Explorer: An Apollo Astronaut's Journey Through the Material and Mystical Worlds, Revised Edition*. New Jersey: New Page Books, 2008.

Murchie, Guy. *The Seven Mysteries of Life: An Exploration in Science and Philosophy*. Massachusetts: Houghton Mifflin Company, 1981.

Oldfield, Harry. "The Human Energy Field and the Invisible Universe." presented at the 16th Annual Conference of the International Society for the Study of Subtle Energies and Energy Medicine. Boulder, CO. 24 June 2006.

Oschman, James L., and Candace Pert. *Energy Medicine: The Scientific Basis*. China: Churchill Livingstone, 2000.

Peers, Allison E. (trans.). "Dark Night of the Soul: Saint John of the Cross, Doctor of the Church." Accessed March 29, 2015. http://www.jesus-passion.com/DarkNightSoulBookI.htm#CHAPTER X.

Pert, Candace B., and Deepak Chopra. *Molecules of Emotion: The Science Behind Mind-Body Medicine*. New York: Touchstone, 1999.

Pew Research Center. "Many Americans Mix Multiple Faiths." Accessed November 23, 2014. http://www.pewforum.org/2009/12/09/many-americans-mix-multiple-faiths.

Radin, Dean. *Entangled Minds: Extrasensory Experiences in a Quantum Reality*. New York: Paraview Pocket Books, 2006.

Schwartz, Gary E., and William L. Simon. *The G.O.D. Experiments: How Science Is Discovering God in Everything, Including Us.* New York: Atria Books, 2006.

Schwartz, Gary E., William L. Simon, and Deepak Chopra. *The Afterlife Experiments: Breakthrough Scientific Evidence of Life after Death.* New York: Atria Books, 2003.

Siegel, Irene R. "Therapist as a Container for Spiritual Resonance and Client Transformation in Transpersonal Psychotherapy: An Exploratory Heuristic Study." *Journal of Transpersonal Psychology* 45, no. 1 (2013): 49–74.

Silva, Freddy. *Secrets in the Fields: The Science and Mysticism of Crop Circles.* Virginia: Hampton Roads Publishing Company, 2002.

Simon, David, and Deepak Chopra. *Return to Wholeness: Embracing Body, Mind, and Spirit in the Face of Cancer.* Canada: John Wiley & Sons, Inc., 1999.

Smolin, Lee. *Three Roads to Quantum Gravity.* New York: Basic Books, 2001.

Sorokin, Pitirim A. *The Ways and Power of Love: Types, Factors, and Techniques of Moral Transformation.* Pennsylvania: Templeton Foundation Press, 2002.

Stevenson, Ian. *Where Reincarnation and Biology Intersect.* Connecticut: Praeger Publishers, 1997.

Swanson, Claude. *The Synchronized Universe: New Science of the Paranormal.* Arizona: Poseidia Press, 2003.

———. *Life Force, the Scientific Basis: Breakthrough Physics of Energy Medicine, Healing, Chi, and Quantum Consciousness; Volume II of the Synchronized Universe Series.* Arizona: Poseidia Press, 2010.

Talbot, Michael. *The Holographic Universe.* New York: Harper Perennial, 1992.

Tart, Charles T. "States of Consciousness." *The Psychedelic Library*. Accessed April 13, 2014. http://www.psychedelic-library.org/soc1.htm.

Tibika, Francoise. *Molecular Consciousness: Why the Universe Is Aware of Our Presence*. Vermont: Park Street Press, 2013.

Tiller, William A., Walter E. Dibble, Jr., and J. Gregory Fandel. *Some Science Adventures with Real Magic*. California: Pavior Publishing, 2005.

Van Praagh, James. *Ghosts Among Us: Uncovering the Truth about the Other Side*. New York: HarperCollins Publishers, 2008.

Walsh, Roger, and Frances Vaughn. *Paths Beyond Ego: The Transpersonal Vision*. New York: Tarcher, 1993.

Weiss, Brian L. *Many Lives, Many Masters: The True Story of a Prominent Psychiatrist, His Young Patient, and the Past-Life Therapy That Changed Both Their Lives*. New York: Fireside, 1988.

Xu, Mingtang. ZY Qigong Retreat with Grandmaster Mingtang Xu: Level I–II. Arizona: June 2014.

Zondervan. *The NIV Study Bible*. Michigan: The Zondervan Corporation, 1985.

Zukav, Gary. *The Dancing Wu Li Masters: An Overview of the New Physics*. New York: A Bantam New Age Book, 1980.

About the Author

Valerie Varan, MS, LPC, NCC is a nationally certified and licensed professional counselor in Colorado. In her holistic, integrative (Eastern–Western psychology) private practice, she sees individuals and couples, specializing in issues of subtle energy awareness and awakening to higher consciousness love.

Made in the USA
Middletown, DE
13 October 2017